STATISTICA

Quick Reference

Chapter

Appendix

STATISTICA Quick Reference

Table of Contents

CHAPTER

STATISTICA: A GENERAL OVERVIEW OF FEATURES

CHAPTER

STATISTICA: A GENERAL OVERVIEW OF FEATURES

STATISTICA is a comprehensive analytic, research, and business intelligence tool. It is an integrated data management, analysis, mining, visualization, and custom application development system featuring a wide selection of basic and advanced analytic procedures for business, data mining, science, and engineering applications.

Analytic Facilities

STATISTICA includes not only general purpose analytic, graphical, and database management procedures, but also comprehensive implementations of specialized methods for data analysis (e.g., predictive data mining, business, social sciences, biomedical research, or engineering applications). All analytic tools offered in the *STATISTICA* line of software are available as part of an integrated package. These tools can be controlled through a selection of alternative user interfaces including:

- a highly optimized interactive user interface (with options to execute *STATISTICA* from within Microsoft Office and other applications),
- a complete thin-client, browser-based user interface (in *WebSTATISTICA*) that enables you to access *STATISTICA* from any computer in the world connected to the Internet, and
- a comprehensive, industry standard, and .NET-compatible programming interface (including the built in, .NET-compatible Visual Basic), offering access to more than 13,000 externally callable functions.

Interactive user interfaces can be easily automated via macros and customized using a variety of methods, and they are recordable in the form of industry standard VB scripts. The built-in development environment can be used to interface *STATISTICA* with other applications and enterprise-wide infrastructures or to build custom extensions of any complexity, from simple shortcuts to advanced, large-scale development projects.

Unique Features

Some of the unique features of the *STATISTICA* line of software include:

- the breadth of selection and comprehensiveness of implementation of analytical procedures,
- the unparalleled selection, quality, and customizability of graphics integrated seamlessly with every computational procedure,
- the efficient and user-friendly user interface,
- the truly open architecture compatible with virtually all enterprise and development environments (including .NET), that exposes *STATISTICA*'s more than 13,000 functions,
- a wide selection of advanced software technologies (see *Software Technology*, below) that is responsible for *STATISTICA*'s practically unlimited capacity, performance (speed, responsiveness), and application customization options.

One of the most unique and important features of the *STATISTICA* family of applications is that these technologies enable even inexperienced users to tailor *STATISTICA* to their specific preferences. You can customize practically every aspect of *STATISTICA,* including even the low-level procedures of its user interface. The same version of *STATISTICA* can be used:

- By novices to perform routine tasks using the default analysis Startup dialog ***Quick*** tab (containing just a few, self-explanatory buttons), or even by accessing *STATISTICA* with their Web browsers (and a highly simplified "front end"), and
- By experienced analysts, professional statisticians, and advanced application developers who can integrate any of *STATISTICA*'s highly optimized procedures (more than 13,000 functions) into custom applications or computing environments, using any of the cutting edge .NET and Web-compatible technologies

The General "Philosophy" of the *STATISTICA* Approach

STATISTICA's default configuration (its general user interface and system options) is a result of years of listening carefully to our users.

We have received feedback from tens of thousands of our users, representing hundreds of thousands of our users from all continents and, practically speaking, "all walks of life." One of the most important facts that we have learned from these users is how different their needs and preferences are (both across individuals and projects or applications). In order to meet those differentiated needs, *STATISTICA* is designed to offer perhaps one of the most flexible and easily customizable user interfaces of any contemporary application.

Although *STATISTICA* provides access to a powerful arsenal of advanced software technologies (see *Software Technology*, below), you do not even need to know about them, because they are designed to work automatically and intuitively. A novice user may never see more than a few self-explanatory buttons. Advanced options, however, are only one tab or mouse click away. Practically every aspect of *STATISTICA* (from the startup configuration, to the way the output is generated and managed by the system, to how *STATISTICA* prompts you to choose your next step) can be changed with a mouse click. Moreover, *STATISTICA* remembers your selections until you change your mind. Practically all dialogs used to select an analysis or perform a routine operation can be easily replaced (e.g., simplified, enhanced, or combined with custom, user-designed procedures). *STATISTICA* will always look and work the way you want.

Software Technology (A Technical Note)

The performance, customizability, and wide selection of options that can be tailored to your needs mentioned in the previous section would not be possible if *STATISTICA* did not feature the advanced technologies that drive all functions of the application.

STATISTICA uses and/or supports virtually all the relevant leading edge software technologies available today. Every one of the more than 13,000 *STATISTICA* functions is accessible to external applications. Practically no limitations are imposed in terms of either the amount or complexity of data that can be stored and

accessed. *STATISTICA* also is optimized for Web and multimedia applications. Computational and graphics procedures are driven by countless proprietary optimizations such as, for example, the "quadruple precision" computational technology that enables us to overcome the limitations of the IEEE floating point storage standards and delivers computational accuracy normally found only in designated math applications (that feature arbitrary-precision options) but not in high volume data processing applications such as statistical or data mining programs.

As a result, *STATISTICA* offers unmatched speed, numerical precision, and responsiveness, which is aided by multithreading (and the advanced "supercomputer-like" distributed/parallel processing architecture offered in the Client-Server version, i.e., *WebSTATISTICA*, and available directly over the Internet).

Data access is based on a flexible streaming technology that enables *STATISTICA* to work effortlessly with both the simple input data files stored on the local drive and queries of multidimensional databases containing terabytes of data and stored in remote data warehouses and processed in-place (i.e., without having to import them to a local storage; this feature is available in enterprise versions of *STATISTICA*).

For example, you can simultaneously run multiple instances of *STATISTICA* [in any combination of local, network, and Client-Server (Web-based) environments], each running multiple analyses of data from multiple and simultaneously open input data files and queries, and the results can be organized into separate projects. *STATISTICA*'s input and output data files and graphs can be of practically unlimited size, comprising hierarchies of documents of various types. The output can be directed to a multitude of output channels such as multimedia tables, high performance workbooks, reports (including *.pdf* files and Microsoft Office documents), and the Internet, as well as the optional *STATISTICA Document Management System*, which can be seamlessly integrated with any *STATISTICA* application.

Web Enablement

One of unique features of the *STATISTICA* family of applications is that it is fully Web enabled, and if the *STATISTICA* Server is installed, you can access the comprehensive functionality of the *STATISTICA* system from any computer in the world connected to the Internet. This includes not merely the options to execute prepared scripts over the Internet, but a plethora of interactive functionality,

including such operations as interactively building predictive data mining models by dragging arrows in the interactive workspace of *STATISTICA Data Miner* (using only the browser, without any client software installed). For more information, please refer to *Appendix B – WebSTATISTICA*, page 269.

Note that most features described in this manual are available in all *STATISTICA* products, although some sections of the manual refer only to specific products such as the *WebSTATISTICA* Server facilities or the *STATISTICA Data Miner* line of products.

Record of Recognition

We are pleased to report that, as of this release, *STATISTICA* has received the highest rating in every published independent comparative review in which it has been featured. In the history of the software industry, very few products have ever achieved such a record.

For more information about StatSoft and *STATISTICA*'s record of recognition, please visit our Web site at ***www.statsoft.com***.

CHAPTER

STEP-BY-STEP EXAMPLES: INTRODUCTORY

CHAPTER

STEP-BY-STEP EXAMPLES: INTRODUCTORY

EXAMPLE 1: CORRELATIONS

Starting *STATISTICA*. After installing *STATISTICA*, you can start the program by selecting ***STATISTICA*** from the Windows ***Start - All Programs*** submenu. You can also double-click on either *STATIST.exe* in Windows Explorer or the icon of any *STATISTICA* file, e.g., a spreadsheet, to start the program.

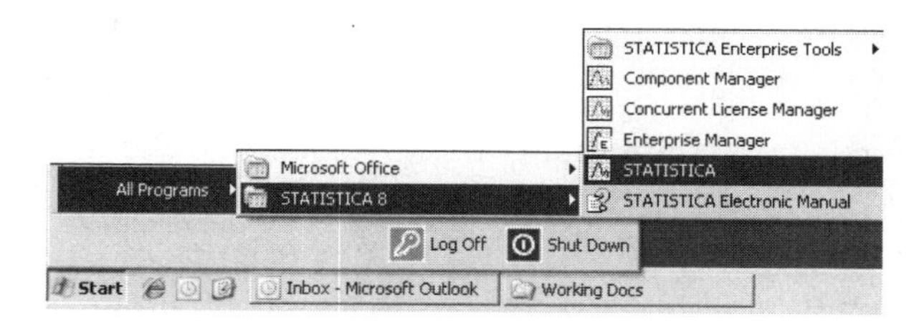

When you start *STATISTICA* for the first time, a blank spreadsheet and the ***Welcome to STATISTICA*** dialog are displayed. The ***Welcome to STATISTICA*** dialog contains options that are useful to access common functions in *STATISTICA*.

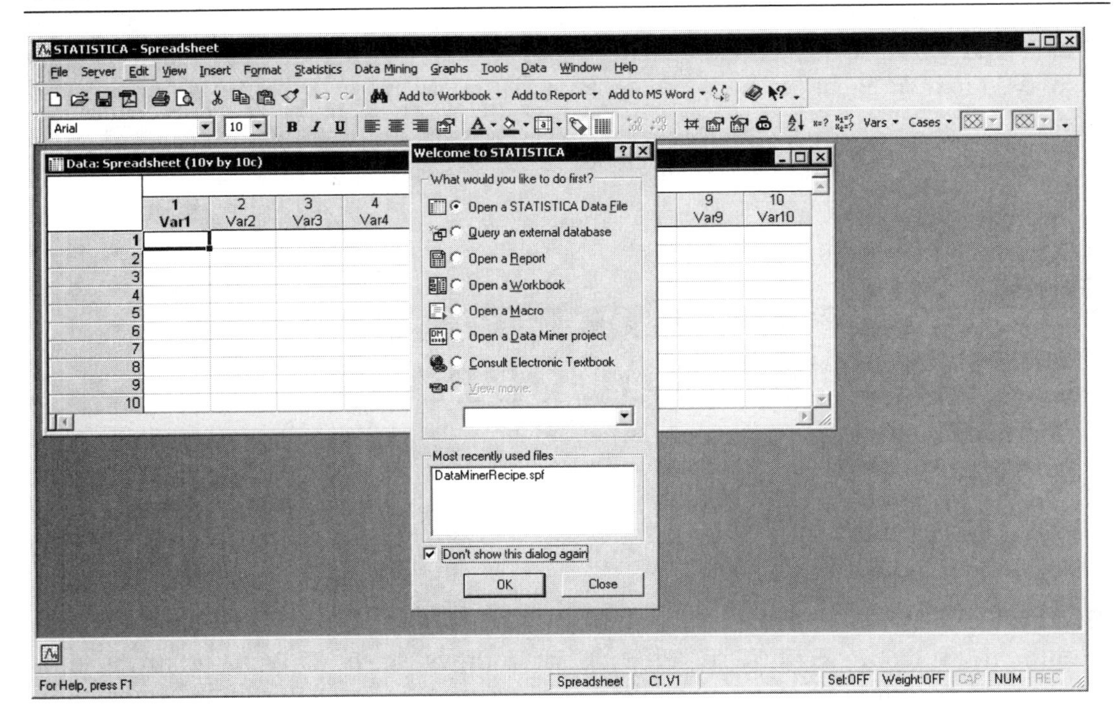

If you prefer, you can select the ***Don't show this dialog again*** check box located near the bottom of the ***Welcome*** dialog, and this dialog will not be displayed when you start *STATISTICA*.

Customization of *STATISTICA*. Note that practically all aspects of the behavior and appearance of *STATISTICA* (even many elementary features illustrated in this example, such as where output is directed) can be permanently customized to match your preferences. For example, even the first step (opening *STATISTICA*) can be customized; you can change the default full-screen opening mode, the appearance of the data spreadsheet, the toolbars, and many other aspects of *STATISTICA*, which will be illustrated throughout this manual.

Selecting a data file. For this example, open *Adstudy.sta*: from the ***File*** menu, select ***Open Examples*** to display the ***Open a STATISTICA Data File*** dialog. Double-click on the *Datasets* folder, and double-click on *Adstudy*. You can also open data files by selecting ***Open*** from the ***File*** menu to display the ***Open*** dialog where you can browse to the appropriate location, click the Open Data button located on each Startup Panel (the first dialog displayed after selecting a command from the ***Statistics***, ***Data-Mining***, or ***Graphs*** menu), or click the toolbar button.

Data spreadsheets (multimedia tables). *STATISTICA* data files are displayed in a spreadsheet (i.e., one spreadsheet is one data file). All *STATISTICA* Spreadsheets are displayed using StatSoft's powerful multimedia table technology, and they can contain not only practically unlimited amounts of data, but also sound, video, embedded documents, automation scripts, and custom user interfaces.

It is possible to have more than one data spreadsheet open at a time (with each spreadsheet connected to a different analysis); thus, most output produced by *STATISTICA* is displayed in spreadsheets (multimedia tables). Note that data management facilities are available from the ***Data*** menu, which is displayed whenever a spreadsheet is open.

The ***Spreadsheet*** toolbar contains the Vars (Variables) and Cases buttons that, when clicked, display menus containing commands to restructure the data file (e.g., ***Add*** or ***Move*** variables or cases).

Vars button menu:

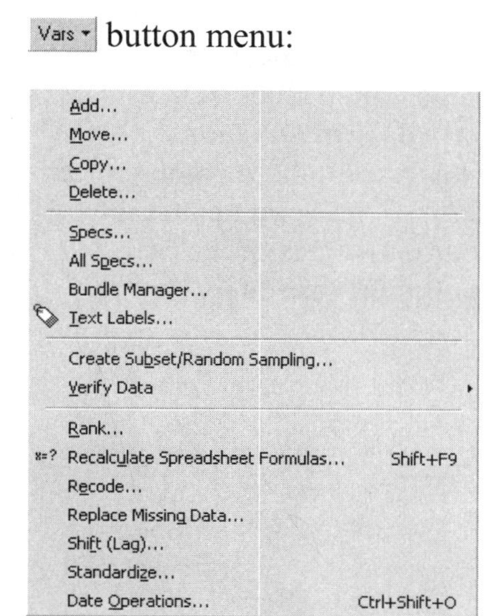

Cases button menu:

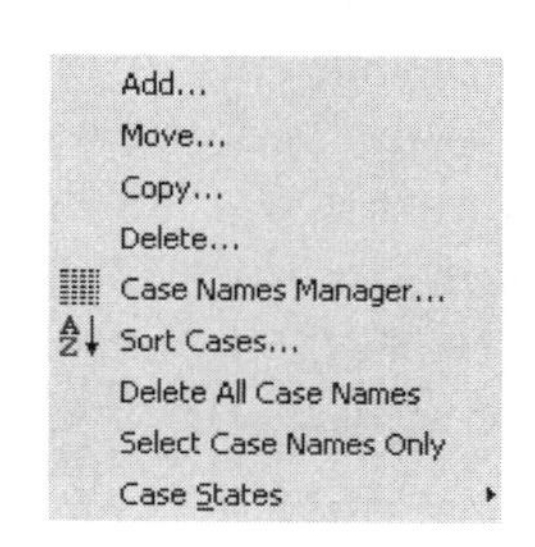

All the commands in the illustrations above are described in the *Electronic Manual*; select (highlight) a command, and press F1 on your keyboard to display the respective Help topic.

Variable specifications. The variable (column) headers in the spreadsheet contain the variable names. Double-click on the first variable header – *GENDER* – to display its ***Variable*** specifications dialog.

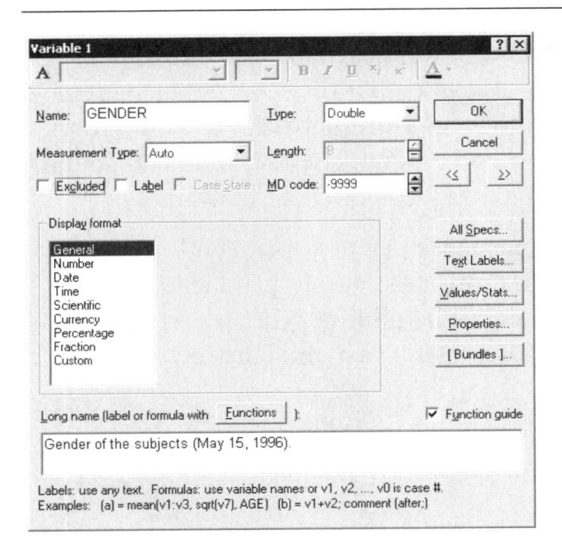

Spreadsheet formulas. Using the options in this dialog, you can change the variable name and/or format, enter a formula to recalculate the values of the variable, etc. If the entry in the ***Long name (label or formula with Functions)*** box starts with an equal sign (**=**), *STATISTICA* interprets it as a formula [a comment can follow after a semicolon (;)]. For example, if you enter into the ***Long name...*** box (of variable one) ***=(v2+v3+v4)/3*** or ***=mean(v2:v4)***, the current values of that variable will be replaced by the average of variables two through four, separately for each case (row) of the spreadsheet.

Specifications of all variables can also be reviewed and edited together in a "combined" ***Variable Specifications Editor*** dialog, accessed by clicking the ***All Specs*** button in the ***Variable*** specifications dialog.

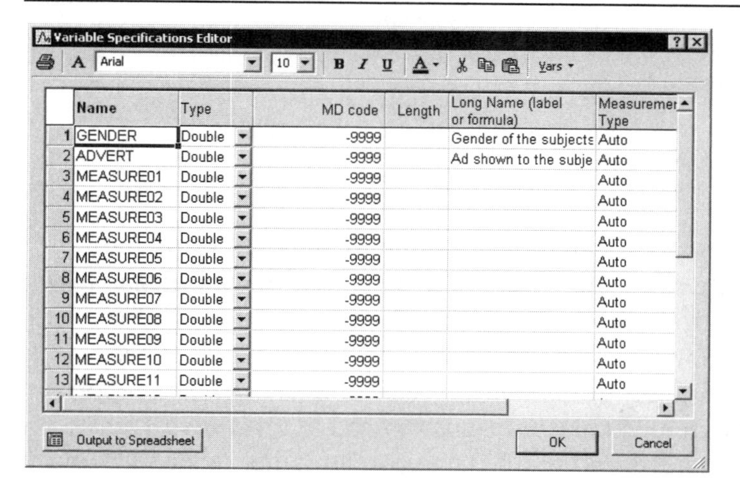

Shortcut menus accessed from spreadsheets. A useful feature of the spreadsheet is the list of commands available from its shortcut menus. Shortcut menus are dynamic menus that are displayed by right-clicking on an item (e.g., a cell in the spreadsheet, as shown in the illustration below). The spreadsheet shortcut menus include a selection of specific data management operations and other options related to the currently selected variable (column), case (row), and/or block of cells.

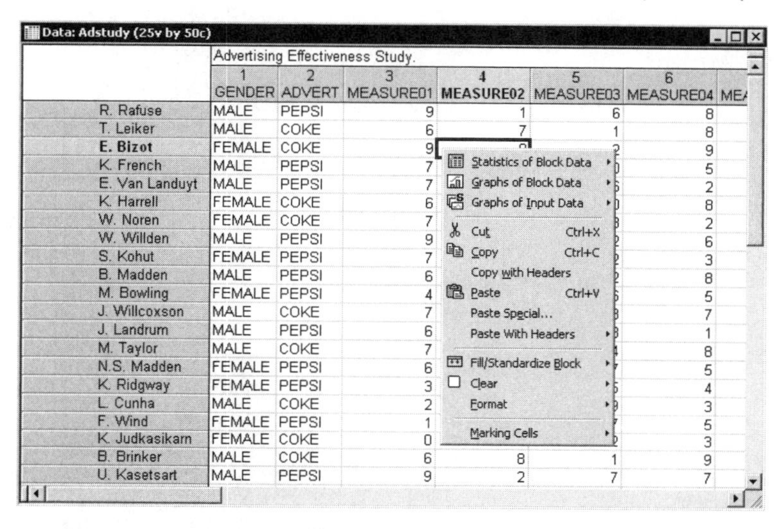

Five ways of handling output. You can customize the way output is managed in *STATISTICA* (see *Five Channels for Output from Analyses*, page 151). You can direct all output to five basic channels:

- Workbooks (see page 152),
- Stand-alone windows (see page 154),
- Reports (see page 155),
- Microsoft Word (see page 158), and
- The Web (see page 160)

The first four output channels listed above are controlled by the options on the ***Output Manager*** tab of the ***Options*** dialog (accessible by selecting ***Output Manager*** from the ***File*** menu or ***Options*** from the ***Tools*** menu). There are a number of ways to output to the Web, depending on the version of *STATISTICA* you have. These means for output can be used in many combinations (e.g., a workbook and report simultaneously), and each output channel can be customized in a variety of ways. Also, all output objects (spreadsheets and graphs) can contain other embedded and linked objects and documents, so *STATISTICA* output can be hierarchically organized in a variety of ways.

Calculating a correlation matrix. Now, let's compute a correlation matrix for the variables in the *Adstudy.sta* data file. To display the ***Basic Statistics and Tables*** Startup Panel, select ***Basic Statistics/Tables*** from the ***Statistics*** menu,

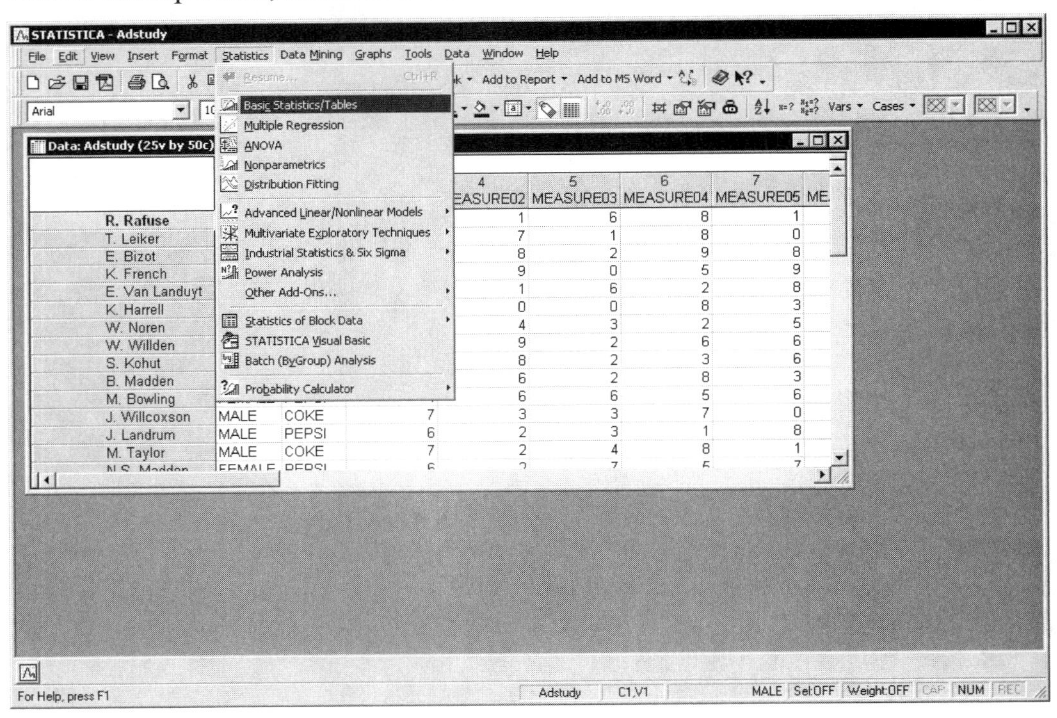

or from the *STATISTICA* Start button menu in the lower-left corner of the screen.

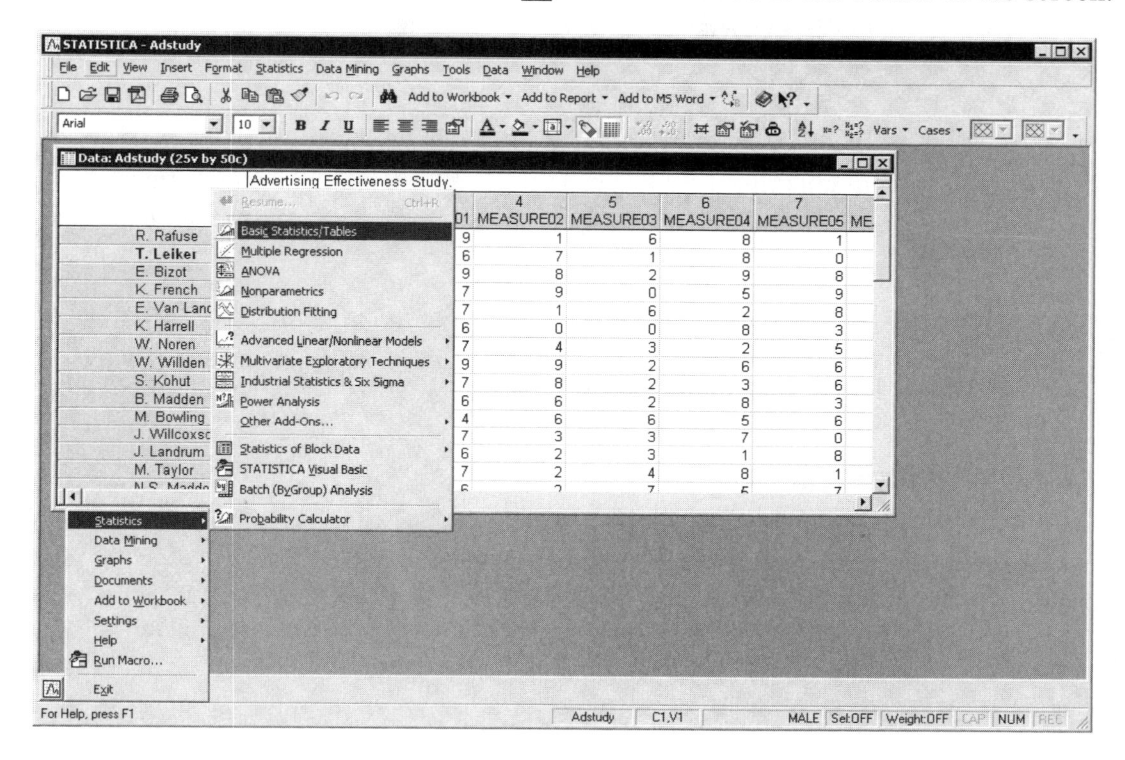

At this point, make sure that a block (a group of selected cells) is *not* selected in the spreadsheet. To deselect a block, click in any cell in the spreadsheet. If a block is selected, *STATISTICA* assumes that the variables corresponding to the block are intentionally preselected for the analysis, and when you later click the ***OK*** or ***Summary*** button to produce the analysis results, instead of prompting you to select variables, *STATISTICA* will automatically produce the correlations for the selected block variables.

In the ***Basic Statistics and Tables*** Startup Panel (shown in the next illustration),

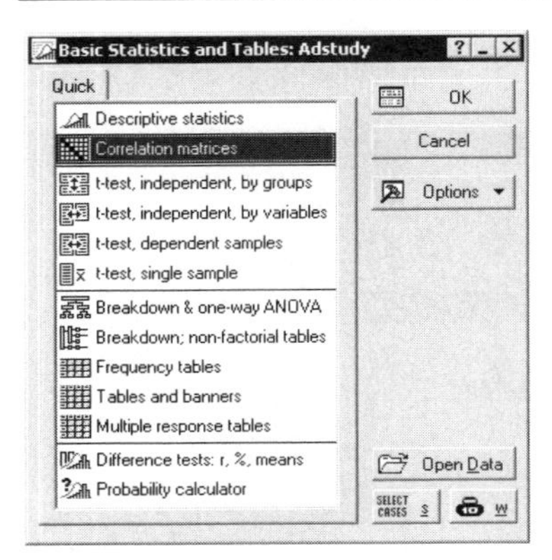

select ***Correlation matrices*** and click the ***OK*** button (or double-click ***Correlation matrices***) to display the ***Product-Moment and Partial Correlations*** dialog.

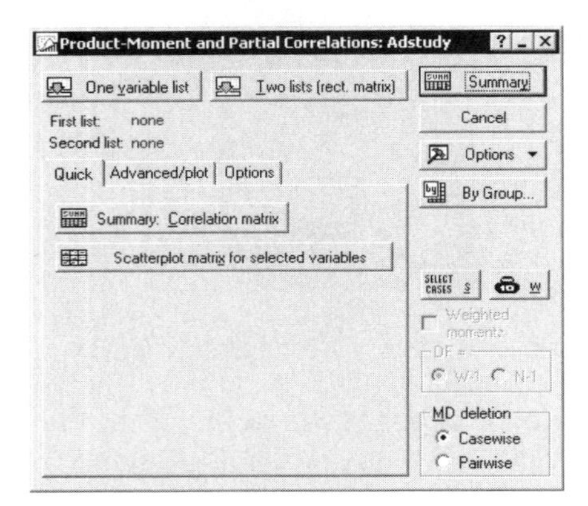

Quick vs. advanced analyses. As with most analysis specification dialogs (and other types of *STATISTICA* dialogs), the ***Product-Moment and Partial Correlations*** dialog is organized by tabs according to the type of options available. Typically, at least two categories of options are available.

The ***Quick*** tab of a dialog contains the most commonly used options, enabling you to quickly specify a basic analysis without having to search through a variety of options.

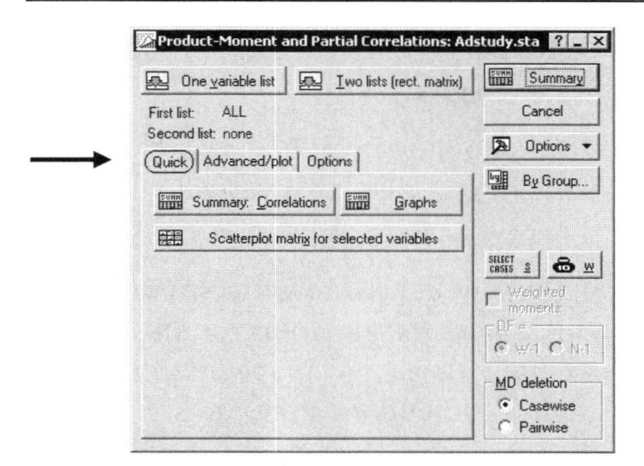

The ***Advanced*** tab typically contains the same options available on the ***Quick*** tab as well as a variety of less commonly used options (e.g., in this case, options to save matrices, produce less commonly requested statistics, and create a variety of plots). Additional tabs are often available as well, depending on the type of analysis being specified.

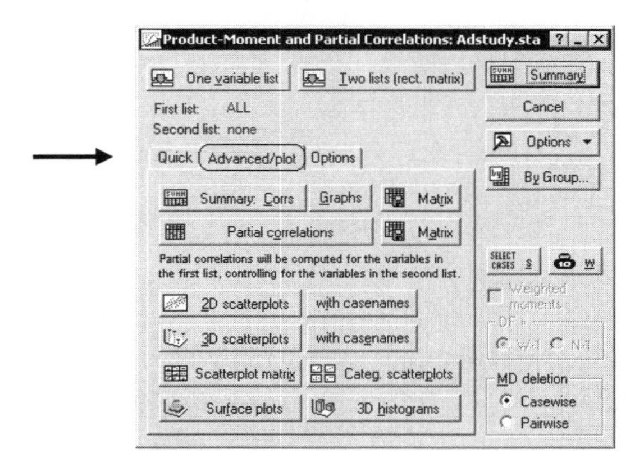

Note that in some cases, only a ***Quick*** tab is available. As with all dialogs in *STATISTICA*, you can press F1 or click the **?** button in the upper-right corner to display a Help topic containing information about the options available on the currently selected tab.

The "self-prompting" nature of *STATISTICA* dialogs. All dialogs in *STATISTICA* follow the "self-prompting" dialog convention, which means that whenever you are not sure what to select next, simply click the ***OK*** button or the

Summary button and *STATISTICA* will proceed to the next logical step, prompting you for the specific input needed (e.g., variables to be analyzed).

Variables button. Every analysis specification dialog in *STATISTICA* contains one or more ***Variable*** buttons that, when clicked, display a variable selection dialog used to specify variables to be analyzed.

Variable selection dialog. For this example, click the ***One variable list*** button (or press ALT+V on your keyboard) to display the ***Select the variables for the analysis*** dialog. Note that the variable selection dialog is also displayed if you click the ***Summary*** button before variables are selected. (As mentioned previously, if a block of variables is selected in the data file, those variables will be selected automatically for the analysis, and when you click the ***Summary*** button, a correlation matrix will be produced for the variables selected in the block, not all variables in the data file.)

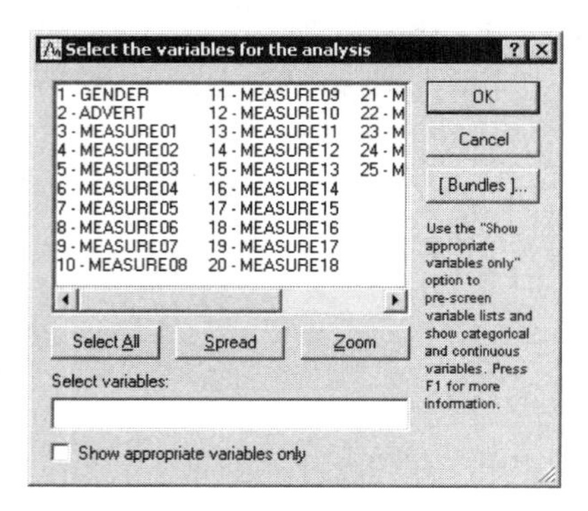

The variable selection dialog supports various ways of selecting variables (including the standard Windows SHIFT+click and CTRL+click conventions to select ranges and discontinuous lists of variables).

You can also use various shortcuts and options in the variable selection dialog to review the contents of the data file. For example, you can spread the variable list to review the variables' long names or formulas (click the ***Spread*** button), or you can zoom in on a variable (click the ***Zoom*** button) to review a sorted list of all values and descriptive statistics for the variable (see the next illustration).

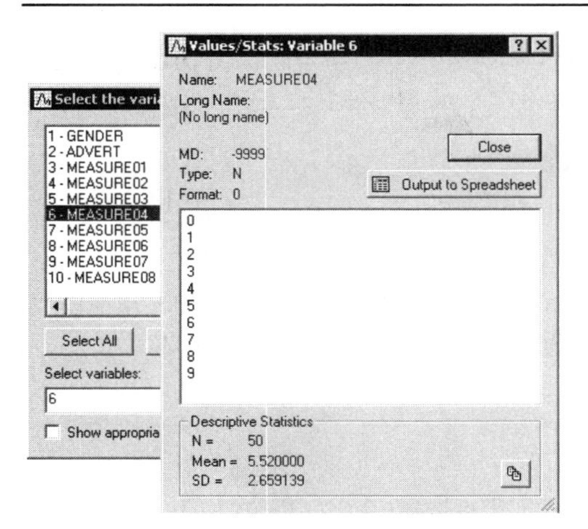

For this example, click the ***Select All*** button in the ***Select the variables for the analysis*** dialog, and then click the ***OK*** button to return to the ***Product-Moment and Partial Correlations*** dialog. Next, click the ***Summary*** button to generate a correlation matrix for the selected variables.

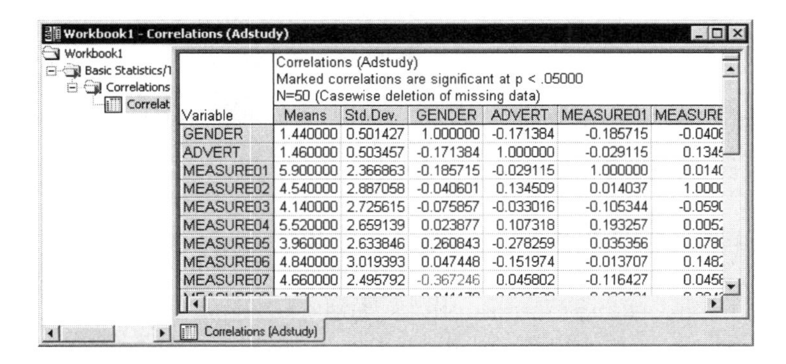

Variable	Means	Std.Dev.	GENDER	ADVERT	MEASURE01	MEASURE
GENDER	1.440000	0.501427	1.000000	-0.171384	-0.185715	-0.0406
ADVERT	1.460000	0.503457	-0.171384	1.000000	-0.029115	0.1345
MEASURE01	5.900000	2.366863	-0.185715	-0.029115	1.000000	0.0140
MEASURE02	4.540000	2.887058	-0.040601	0.134509	0.014037	1.0000
MEASURE03	4.140000	2.725615	-0.075857	-0.033016	-0.105344	-0.0590
MEASURE04	5.520000	2.659139	0.023877	0.107318	0.193257	0.0052
MEASURE05	3.960000	2.633846	0.260843	-0.278259	0.035356	0.0780
MEASURE06	4.840000	3.019393	0.047448	-0.151974	-0.013707	0.1482
MEASURE07	4.660000	2.495792	-0.367246	0.045802	-0.116427	0.0458

Note that instead of clicking the ***Summary*** button, you could have clicked the ***Summary: Correlations*** button on the ***Quick*** tab or on the ***Advanced*** tab. Also, depending on the defaults you have specified for handling output (on the ***Output Manager*** tab of the ***Options*** dialog), the *Correlations* spreadsheet can be displayed in a report or a stand-alone window or sent to a Word document, rather than in a workbook as shown above.

Summary graphs. *STATISTICA* provides extremely flexible tools and methods for summarizing key results in graphs and/or tables. For example, click the

Graphs button to display summary graphs for each pair of variables in the correlation matrix.

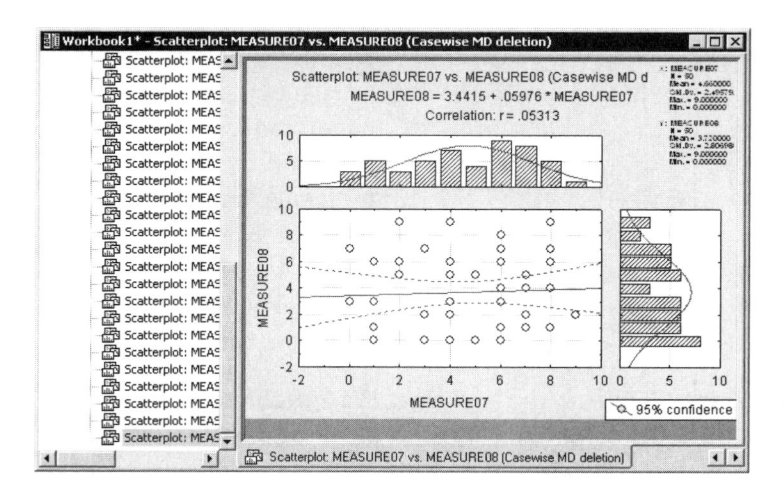

These graphs not only show the scatterplot of points for each correlation, but also the distributions (histograms) for each variable, as well as the respective correlation coefficient and regression equation.

STATISTICA incorporates many such displays to summarize basic descriptive statistics, correlations, the results of Gage or Process capability studies, or other types of data analyses.

Results spreadsheets (multimedia tables). In addition to storing data, spreadsheets are used in *STATISTICA* to display most of the numeric output. Note that spreadsheets offer many display features and options, and in this example, significant correlations are marked with a different format to help distinguish them; by default, the color is red (in the *Correlations* spreadsheet, see the cell adjacent to *MEASURE07*, under *GENDER*). Spreadsheets can hold anywhere from a short line to gigabytes of output, and they offer a variety of options to facilitate reviewing the results and visualizing them in predefined and custom-defined graphs, as will be seen later in this example. Also, as mentioned previously, *STATISTICA* Spreadsheets are managed using StatSoft's powerful multimedia table technology. They can handle not only virtually unlimited amounts of data, but also video, sound, custom user interfaces, and auto-executing scripts, as well as offer virtually unlimited customization options (see page 181 for further details on spreadsheets).

Spreadsheet options. Most spreadsheet facilities are accessible via buttons on the Spreadsheet toolbar and the shortcut menus (displayed by right-clicking in any cell). You can try these options to see how they work, or you can review their descriptions by pressing the help key (F1) or clicking the toolbar button and then clicking on the respective toolbar button. You can change all aspects of the display formats for each spreadsheet column, edit the output, or append blank cases and variables to make room for notes or output pasted from other sources. Spreadsheets can be printed in a variety of ways (by default, in presentation-quality tables with grid lines). Also, since spreadsheets are used for input, you can easily specify an analysis using the results from a previous analysis (for example, you could use this correlation matrix to specify a multidimensional scaling analysis). To use a results spreadsheet as an input spreadsheet, select ***Input Spreadsheet*** from the ***Data*** menu when that spreadsheet is active.

Analysis workbooks and other output options. All results can be displayed (and stored) in stand-alone windows, reports, Word documents, or workbooks, which represent the default (and perhaps the most versatile) way of handling output from analyses (see page 152 and page 177 for further details on workbooks). Depending on your selections in the ***Output Manager*** (accessible by selecting ***Output Manager*** from the ***File*** menu, see the next paragraph), results can be put in a single workbook that holds the results from all analyses, a separate analysis workbook that holds the results (spreadsheets and graphs) from a single analysis, the workbook that contains the original data file, or a preexisting workbook. Additionally, you can choose to have the results sent to a workbook automatically, or you can send them to the workbook yourself by clicking the Add to Workbook toolbar button to send selected stand-alone spreadsheets or graphs to a workbook.

Output Manager. Which type of workbook you choose, or whether you choose to use a workbook, depends entirely on how you prefer to store your data and results. To change the output destination for the results of a particular analysis only, click the Options button on any analysis or graph specification dialog, and select ***Output*** to display the ***Analysis/Graph Output Manager*** dialog.

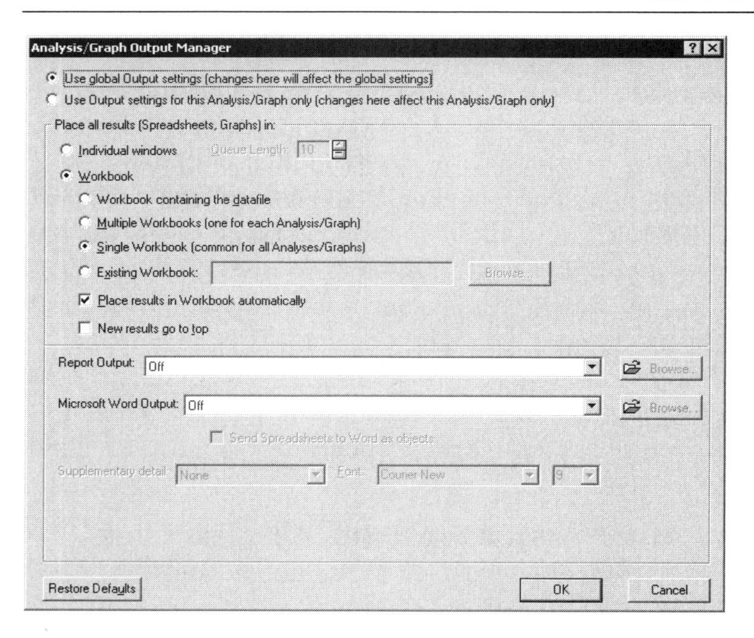

To change output options for all analyses, use the (global) ***Output Manager*** (the ***Output Manager*** tab of the ***Options*** dialog, accessible by selecting ***Output Manager*** from the ***File*** menu or ***Options*** from the ***Tools*** menu), or select the ***Use global Output settings (changes here will affect the global settings)*** option button on the ***Analysis/Graph Output Manager*** dialog.

As with all workbooks, individual documents (e.g., spreadsheets or graphs) or groups of documents can be printed, extracted, copied, and deleted from an analysis workbook. See the overview of *Workbooks* on page 177 for more details; see also the *Electronic Manual*.

Copy vs. Copy with Headers. Contents of spreadsheets can be copied to the Clipboard by pressing CTRL+C (which copies only the contents of the selected block) or by selecting ***Copy with Headers*** from the ***Edit*** menu (which copies the block along with its respective variable and case names). If pasted into a word processor document, spreadsheets will appear as active (in-place editable) *STATISTICA* objects, standard RTF-formatted tables, unformatted text, pictures, or HTML (depending on your choice in the ***Paste Special*** dialog of the word processor).

Printing spreadsheets. To produce a hard copy of the output spreadsheets, select ***Print*** from the ***File*** menu (or press CTRL+P) to display the ***Print*** dialog, in

which you specify printing options. You can also use the shortcut method of clicking the printer toolbar button. This shortcut method does not display the ***Print*** dialog, but prints the entire current document. If you want to print a document from within a workbook, make sure the document is selected in the workbook and select the ***Selection*** option button on the ***Print*** dialog. You can also extract a copy of the document from the workbook (drag it from the tree pane, or select the document and select ***Extract as stand-alone window*** from the ***Workbook*** menu) and then print it.

Optional reports of all output. Workbooks offer perhaps the most flexible options to manage your output (see pages 152 and 177). In some circumstances, however, it may be useful to automatically produce a log of all results (contents of all spreadsheets and/or graphs) in a traditional word processor style report format where comments and annotations can be inserted in arbitrary locations, objects can be placed side by side, etc. (see page 154 and page 185 for further details on reports).

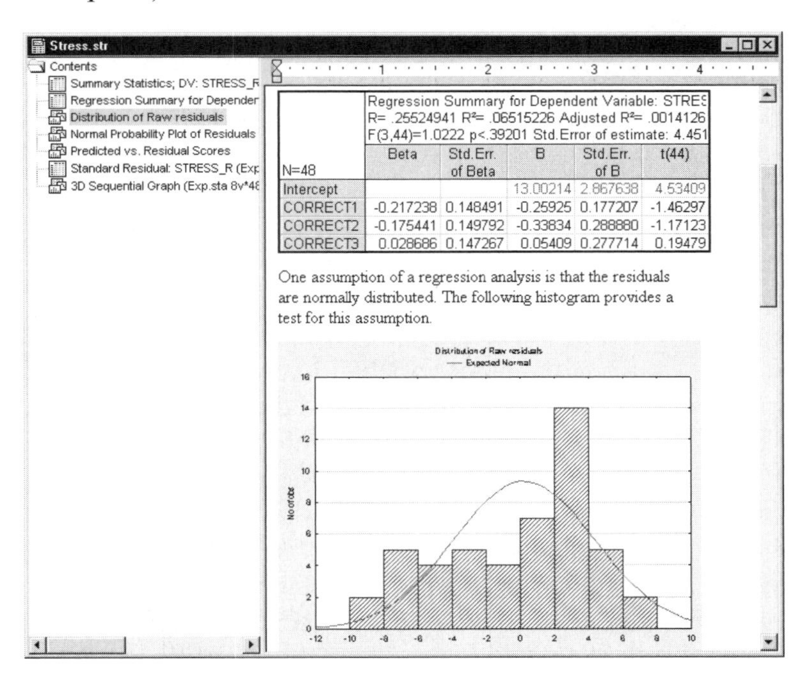

N=48	Beta	Std.Err. of Beta	B	Std.Err. of B	t(44)
Intercept			13.00214	2.867638	4.53409
CORRECT1	-0.217238	0.148491	-0.25925	0.177207	-1.46297
CORRECT2	-0.175441	0.149792	-0.33834	0.288880	-1.17123
CORRECT3	0.028686	0.147267	0.05409	0.277714	0.19479

In order to create such a report, on the ***Output Manager*** tab of the ***Options*** dialog or on the ***Analysis/Graph Output Manager*** dialog, click the arrow adjacent to ***Report Output***. From the drop-down menu, select either ***Send to Multiple Reports (one for***

each Analysis/Graph), ***Single Report (common for all Analyses/graphs)***, or ***(Select File)*** (which will display the ***Open*** dialog where you can select an already established report). To display the ***Output Manager***, select ***Output Manager*** from the ***File*** menu or ***Options*** from the ***Tools*** menu and click on the ***Output Manager*** tab (for global changes). To display the ***Analysis/Graph Output Manager*** dialog, click the Options button on any analysis or graph specification dialog and select ***Output*** (for local changes). Using the ***Output Manager***, you can also specify the amount of supplementary information to be included with the spreadsheet results.

Interpretation of the results (Electronic Statistics Textbook). Now let's return to the example and the correlation matrix that has been produced.

Workbook1 - Correlations (Adstudy)

Correlations (Adstudy)
Marked correlations are significant at $p < .05000$
N=50 (Casewise deletion of missing data)

Variable	Means	Std.Dev.	GENDER	ADVERT	MEASURE01	MEASURE
GENDER	1.440000	0.501427	1.000000	-0.171384	-0.185715	-0.040
ADVERT	1.460000	0.503457	-0.171384	1.000000	-0.029115	0.134
MEASURE01	5.900000	2.366863	-0.185715	-0.029115	1.000000	0.014
MEASURE02	4.540000	2.887058	-0.040601	0.134509	0.014037	1.000
MEASURE03	4.140000	2.725615	-0.075857	-0.033016	-0.105344	-0.059
MEASURE04	5.520000	2.659139	0.023877	0.107318	0.193257	0.005
MEASURE05	3.960000	2.633846	0.260843	-0.278259	0.035356	0.078
MEASURE06	4.840000	3.019393	0.047448	-0.151974	-0.013707	0.148
MEASURE07	4.660000	2.495792	-0.367246	0.045802	-0.116427	0.045

Correlations (Adstudy)

Each of the cells of the correlation matrix represents a value (in the range of –1.00 to +1.00) that reflects the relation between the variables (see the respective variable and case headers). The higher the absolute value of the correlation coefficient, the closer the relation; if the value is positive, the relation is "positive" (high values of one variable correspond to high values of the other variable; likewise, low values of one variable correspond to low values of the other variable). If the value is negative, the opposite is true (low values of one variable correspond to high values of the other variable). To learn more about how to interpret values of correlations, you can review a comprehensive, illustrated discussion of the topic in the *Electronic Manual*, which features the complete contents of the StatSoft *Electronic Statistics Textbook* (an award-wining, Web-based general resource on statistics that has been recommended by *Encyclopedia Britannica* for its "Quality, Accuracy, Presentation, and Usability"). See also, the ***Statistical Advisor*** (page 33).

To display the *Electronic Manual*, select ***STATISTICA Help*** from the ***Help*** menu. On the ***Search*** tab of the *Electronic Manual*, enter the respective term (e.g., ***Correlations***) into the ***Type in the word(s) to search for*** box, click the ***List Topics*** button, and then select the desired topic in the ***Select topic*** box (in this case ***Correlations - Introductory Overview***):

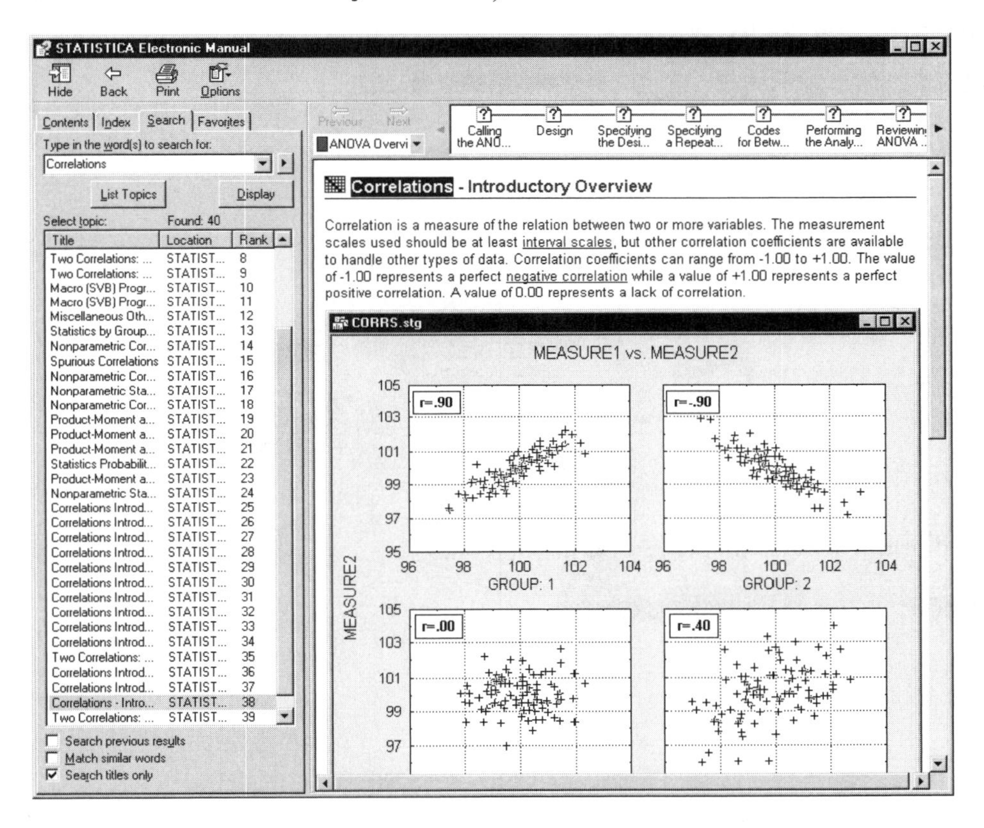

One of the important (and often overlooked) issues discussed in the *Electronic Manual* is the importance of scatterplots in examining correlations. For example, even very large and highly statistically significant correlation coefficients can be entirely due to one unusual data point ("outlier"), and if that is the case, then the correlation coefficient (even if statistically significant) would have no value to us (i.e., it would have no "predictive validity"). Following this concern, and the advice of the *Electronic Statistics Textbook*, let's examine a scatterplot that will visualize a relation between the variables and, thus, visualize a particular correlation coefficient from the table.

Producing graphs from spreadsheets. While examining the spreadsheet, you can view the correlations graphically, for example, to visualize the correlation between variables ***Measure09*** and ***Measure05***. To produce a scatterplot for these two variables, right-click on the respective correlation coefficient (***-.467199***). In the resulting shortcut menu, select ***Graphs of Input Data - Scatterplot by MEASURE09 - Regression, 95% conf.***, as shown below.

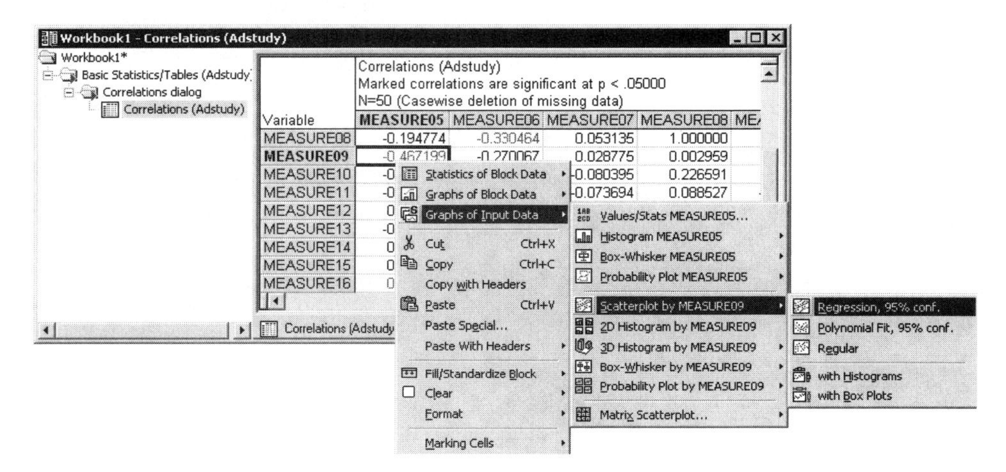

The specified graph will be displayed on the screen.

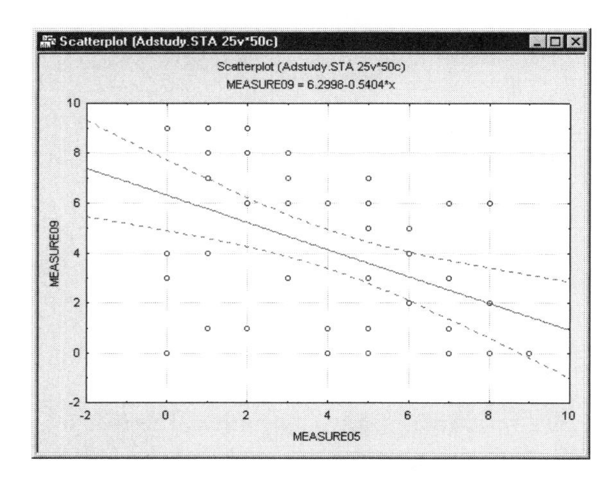

As we can learn from the graph, there are no unusual patterns of data, thus, there is no reason to be concerned about outliers (see the short discussion of outliers on page 27; see also the topic on outliers in the *Electronic Manual*).

Graph customization. Note that now, when the focus is on the graph window, the toolbar has changed. The Graph Tools toolbar (which accompanies all graph windows) looks different from the toolbar for the spreadsheets.

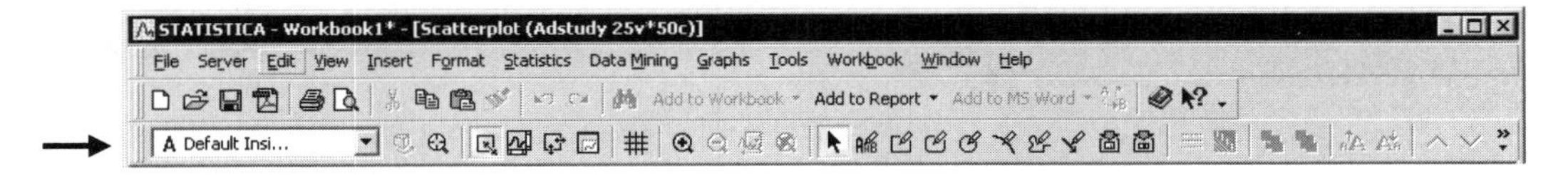

It contains a variety of graph customization and drawing tools. All of these options are also available from menus, and most of them are available from shortcut menus accessed by right-clicking on specific parts of the graph. Note that the options on shortcut menus are hierarchical, meaning that the first one or two options apply specifically to the graph element you have selected, while lower options will display dialogs that offer more options on a greater variety of graph elements related to the element you have selected. If you right-click anywhere n the space outside the graph axes, a menu of global options is displayed (as shown below).

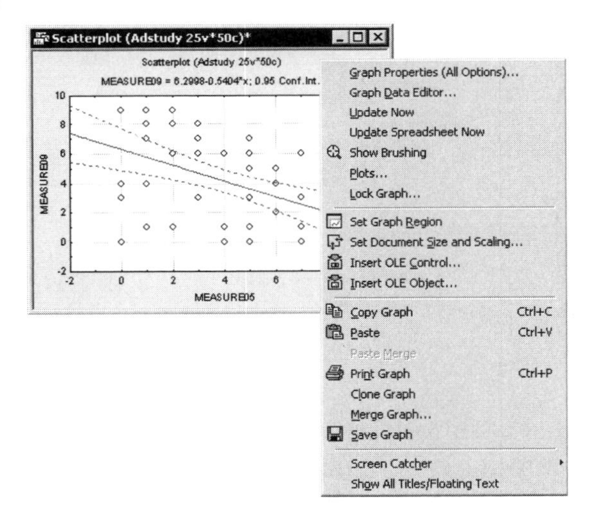

For more information on graph customization, see page 198, and the *Electronic Manual*.

Now let's return to the spreadsheet.

Split scrolling in spreadsheets. Spreadsheets can be split into up to four sections (panes) by dragging the split box (the small rectangle at the top of the vertical scrollbar or to the left of the horizontal scrollbar). This is useful if you have

a large amount of information and you want to review results from different parts of the spreadsheet. When you move the mouse pointer to the split box, the mouse pointer changes to +||+ or ÷. Now, to position the split, drag it to the desired position.

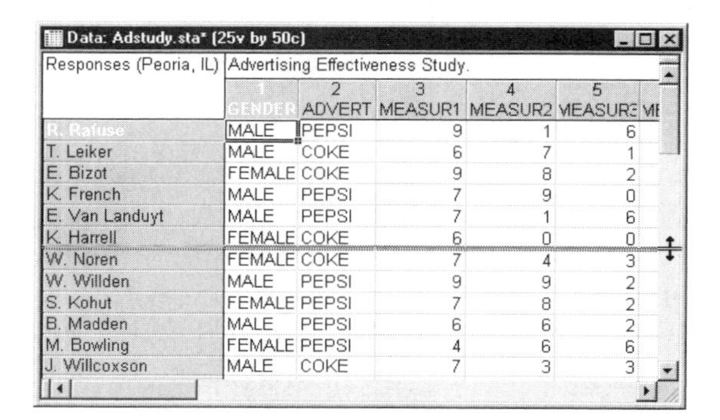

Data: Adstudy.sta* (25v by 50c)

Responses (Peoria, IL)	1 GENDER	2 ADVERT	3 MEASUR1	4 MEASUR2	5 MEASUR3
R. Rafuse	MALE	PEPSI	9	1	6
T. Leiker	MALE	COKE	6	7	1
E. Bizot	FEMALE	COKE	9	8	2
K. French	MALE	PEPSI	7	9	0
E. Van Landuyt	MALE	PEPSI	7	1	6
K. Harrell	FEMALE	COKE	6	0	0
W. Noren	FEMALE	COKE	7	4	3
W. Willden	MALE	PEPSI	9	9	2
S. Kohut	FEMALE	PEPSI	7	8	2
B. Madden	MALE	PEPSI	6	6	2
M. Bowling	FEMALE	PEPSI	4	6	6
J. Willcoxson	MALE	COKE	7	3	3

You can change the position of the split by dragging the split box (now located between panes) to a new position.

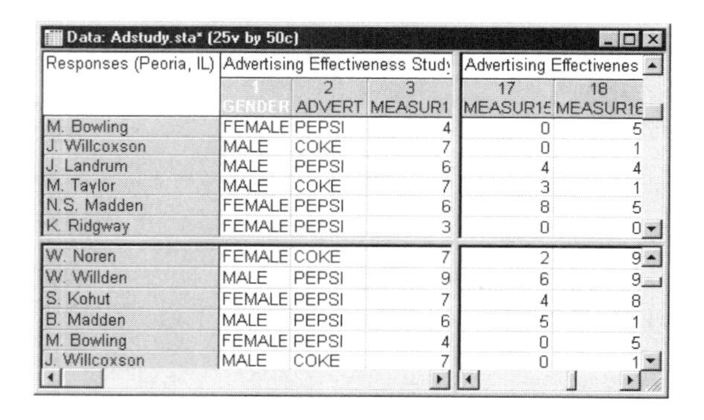

Data: Adstudy.sta* (25v by 50c)

Responses (Peoria, IL)	1 GENDER	2 ADVERT	3 MEASUR1	17 MEASUR15	18 MEASUR16
M. Bowling	FEMALE	PEPSI	4	0	5
J. Willcoxson	MALE	COKE	7	0	1
J. Landrum	MALE	PEPSI	6	4	4
M. Taylor	MALE	COKE	7	3	1
N.S. Madden	FEMALE	PEPSI	6	8	5
K. Ridgway	FEMALE	PEPSI	3	0	0
W. Noren	FEMALE	COKE	7	2	9
W. Willden	MALE	PEPSI	9	6	9
S. Kohut	FEMALE	PEPSI	7	4	8
B. Madden	MALE	PEPSI	6	5	1
M. Bowling	FEMALE	PEPSI	4	0	5
J. Willcoxson	MALE	COKE	7	0	1

Note that vertically split panes scroll together when you scroll horizontally; horizontally split panes scroll together when you scroll vertically. For information about highlighting blocks of data across split panes and about variable-speed highlighting of blocks of data, see *How can I expand a block in the spreadsheet outside the current screen?* in the *Electronic Manual*.

Drag-and-drop. *STATISTICA* supports the complete set of standard spreadsheet (Microsoft Excel-style) drag-and-drop facilities. For example, in order to move a block, point to the border of the selection (the mouse pointer changes to an arrow) and drag it to the new location.

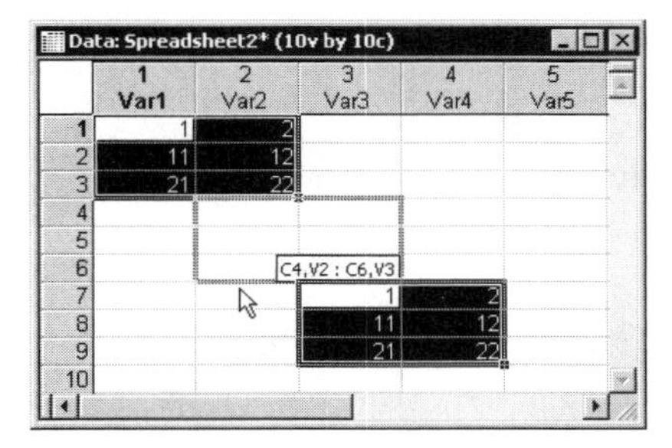

To copy a block of data, point to the border of the selection (the mouse pointer changes to an arrow) then drag the selection to a new location while pressing the CTRL key. Note that when you are dragging the selection, a plus sign (+) is displayed next to the mouse pointer to indicate you are copying the text rather than moving it (see the image, below).

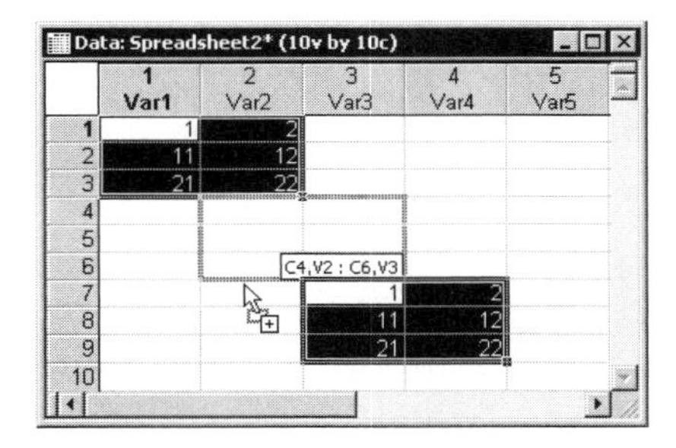

To insert a block between columns or rows, point to the border of the selection (the mouse pointer changes to an arrow) and then drag the selection while pressing the SHIFT key.

If you point between rows, an insertion bar is displayed between the rows, and when you release the mouse button, the block is inserted between those two rows [creating new case(s)]. If you point between columns, an insertion bar is displayed between the columns, and when you release the mouse button, the block is inserted between those two columns [creating new variable(s)].

Note that if you also press the CTRL key while you are dragging the selection, the block will be copied and inserted instead of moved and inserted; a plus will appear next to the mouse pointer (as shown in the next illustration).

Additionally, a series of values within a block can be extrapolated (AutoFilled) by dragging the Fill Handle (a small, solid square located on the lower-right corner of the block border).

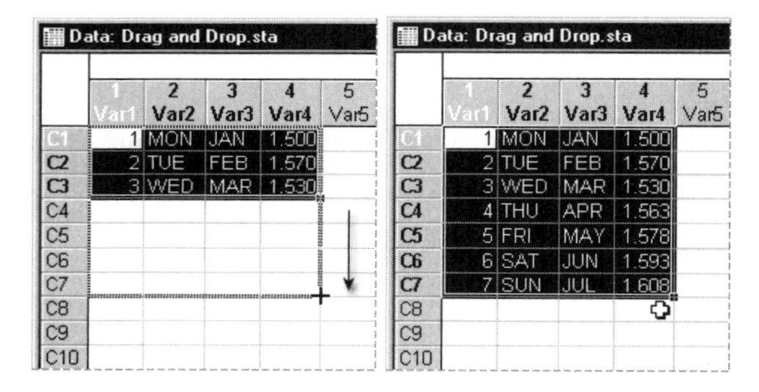

Electronic Manual. *STATISTICA* provides an *Electronic Manual* with comprehensive documentation on all program procedures and all options, available in a context-sensitive manner (there is a total of more than 100 megabytes of compressed documentation included). To access the manual, select ***STATISTICA Help*** from the ***Help*** menu or click the button on the *STATISTICA* toolbar. You can also select (highlight) a menu command or select a tab for which you want information, and press F1 on your keyboard to display the respective Help topic, or click the help button that is on the caption bar of all dialogs.

Due to its dynamic hypertext organization, organizational tabs (***Contents***, ***Index***, ***Search***, and ***Favorites***), and various facilities used to customize the help system, it is faster to use the *Electronic Manual* than to look for information in the traditional manuals.

The status bar at the bottom of the *STATISTICA* window also displays short explanations of the menu commands or toolbar buttons when an item is selected or a button is clicked.

Statistical Advisor. A ***Statistical Advisor*** facility is built into the *STATISTICA Electronic Manual*. Select ***Statistical Advisor*** from the ***Help*** menu to display a set of simple questions about the nature of the research problem and the type of your data. Click on the appropriate links to answer the questions, and suggestions for the statistical procedures that appear most relevant will be displayed, containing links to guide you to the specific procedures in the *STATISTICA* system.

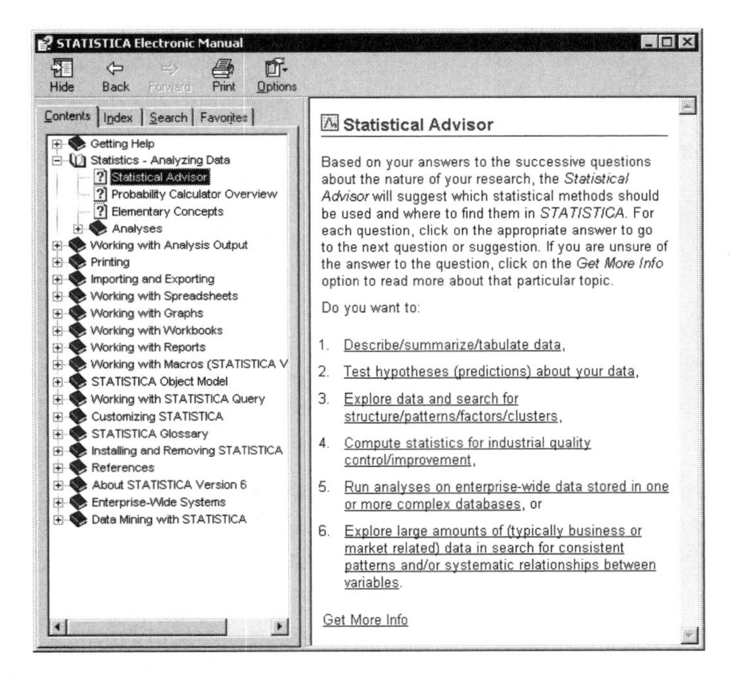

Direct jumps (hypertext links) in the ***Statistical Advisor*** topics guide you to corresponding *Introductory Overviews*, which discuss in detail the respective statistical methods and procedures.

EXAMPLE 2: ANOVA

Calling the ANOVA module. For this example of a 2 x 2 (between) x 3 (repeated measures) design, open the *Adstudy.sta* data file. Then, to start the ***ANOVA/MANOVA*** analysis, select ***ANOVA*** from the ***Statistics*** menu to display the ***General ANOVA/MANOVA*** Startup Panel.

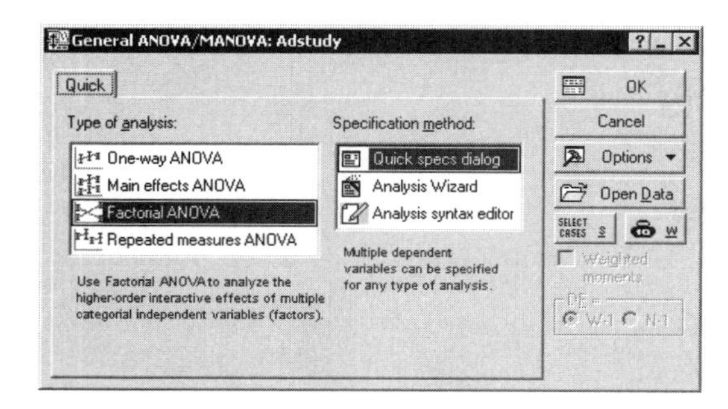

This dialog is used to specify very simple analyses (e.g., via ***One-way ANOVA*** – designs with only one between-group factor) and more complex analyses (e.g., via ***Repeated measures ANOVA*** – designs with between-group factors and a within-subject factor).

Design. Select ***Repeated measures ANOVA*** as the ***Type of analysis*** and ***Quick specs dialog*** as the ***Specification method***, and then click the ***OK*** button in the ***General ANOVA/MANOVA*** Startup Panel to display the ***ANOVA/MANOVA Repeated Measures ANOVA*** dialog.

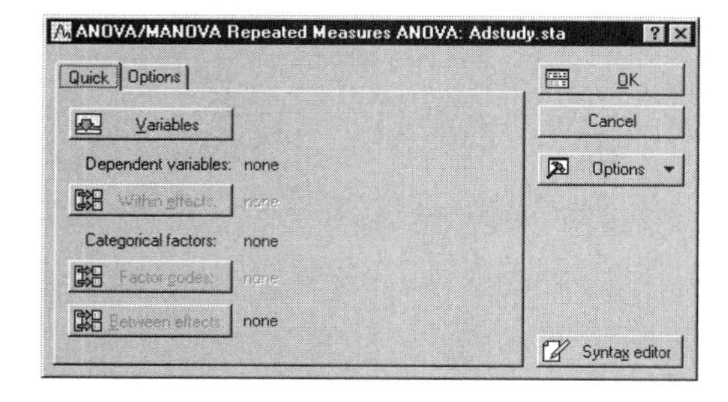

Specifying the design (variables). The first (between-group) factor is ***Gender*** (with 2 levels: ***Male*** and ***Female***). The second (between-group) factor is ***Advert*** (with 2 levels: ***Pepsi*** and ***Coke***). The two factors are crossed, which means that there are both ***Male*** and ***Female*** subjects in the ***Pepsi*** and ***Coke*** groups. Each of those subjects responded to three questions (this repeated measure factor will be called ***Response***; it has three levels represented by variables ***Measure01***, ***Measure02***, and ***Measure03***).

Click the ***Variables*** button (in the ***ANOVA/MANOVA Repeated Measures ANOVA*** dialog) to display the variable selection dialog. Select *Measure01* through *Measure03* as dependent variables (from the ***Dependent variable list*** field) and *Gender* and *Advert* as factors [from the ***Categorical predictors (factors)*** field].

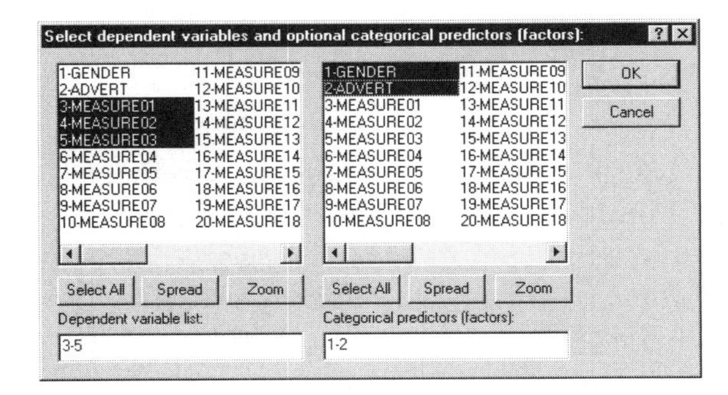

Then click the ***OK*** button to return to the ***ANOVA/MANOVA Repeated Measures ANOVA*** dialog.

The repeated measures design. The design of the experiment that we are about to analyze can be summarized as follows:

	Between-Group Factor #1: *Gender*	Between-Group Factor #2: *Advert*	Repeated Measure Factor: *Response* Level #1: *Measure01*	Level #2: *Measure02*	Level #3: *Measure03*
Subject 1	Male	Pepsi	9	1	6
Subject 2	Male	Coke	6	7	1
Subject 3	Female	Coke	9	8	2
⋮	⋮	⋮	⋮	⋮	⋮

Specifying a repeated measures factor. The minimum necessary selections are now complete, and, if you did not care about selecting the repeated measures factor, you would be ready to click the ***OK*** button and see the results of the analysis. However, for our example, you need to specify that the three dependent variables you have selected be interpreted as three levels of a repeated measures (within-subject) factor. Unless you do so, *STATISTICA* assumes that those are three "different" dependent variables and runs a ***MANOVA*** (i.e., ***Multivariate ANOVA***).

In order to define the desired repeated measures factor, click the ***Within effects*** button on the ***Quick*** tab to display the ***Specify within-subjects factor*** dialog.

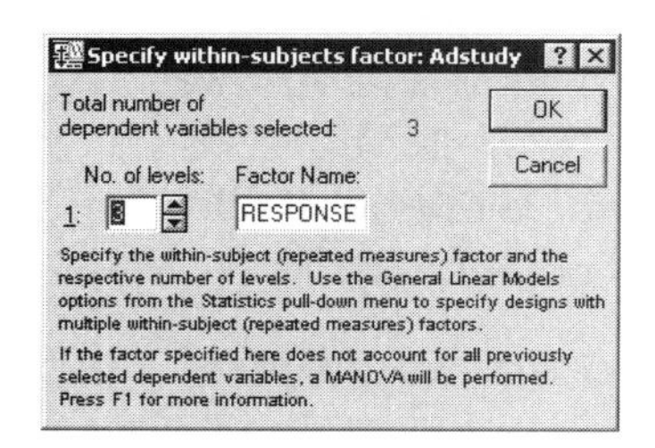

Note that *STATISTICA* has suggested the selection of one repeated measures factor with ***3*** levels (default name ***R1***). You can specify only one within-subject (repeated measures) factor via this dialog. To specify multiple within-subject factors, use the ***General Linear Models*** module (available in the optional ***Advanced Linear/Nonlinear Models*** package). Press the F1 key on your keyboard while the ***Specify within-subjects factor*** dialog is displayed (or click the ? button in the upper-right corner of the dialog) to display an *Electronic Manual* topic containing links to comprehensive discussions of repeated measures and examples of designs.

For this example, edit the name for the factor: in the ***Factor Name*** box, change the default *R1* to *RESPONSE*, and click the ***OK*** button to exit the dialog.

Codes (defining the levels) for between-group factors. You do not need to manually specify codes for between-group factors [i.e., instruct *STATISTICA* that variable *Gender* has two levels: *1* and *2* (or *Male* and *Female*)] unless you want to prevent *STATISTICA* from using, by default, all codes encountered in the selected grouping variables in the data file. To enter such custom code selection, click the ***Factor codes*** button to access the ***Select codes for indep. vars (factors)*** dialog.

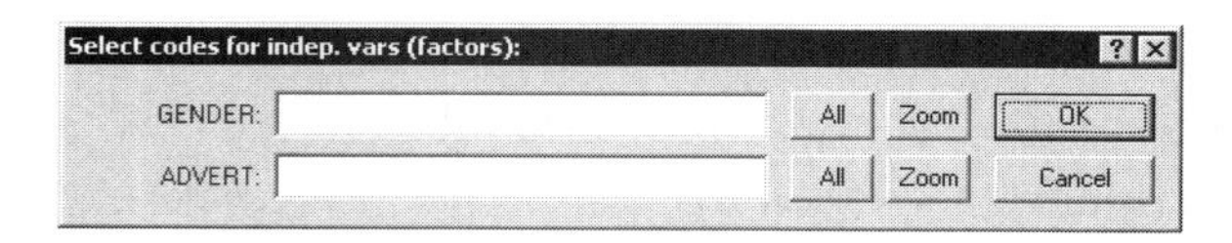

You can use the options in this dialog to review values of individual variables before you make your selections by clicking the ***Zoom*** button, scan the file, and fill in the codes fields (e.g., ***Gender*** and ***Advert***) for an individual variable or all variables, etc. For now, click the ***OK*** button in the ***Select codes for indep. vars (factors)*** dialog; *STATISTICA* automatically fills in the codes fields with all distinctive values encountered in the selected variables,

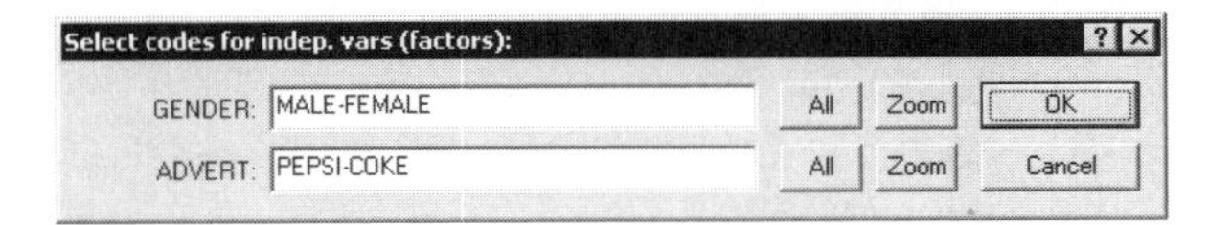

and closes the dialog.

Performing the analysis. Click the ***OK*** button in the ***ANOVA/MANOVA Repeated Measures ANOVA*** dialog. The analysis is performed and the ***ANOVA Results*** dialog is displayed, which contains various output spreadsheets and graphs options.

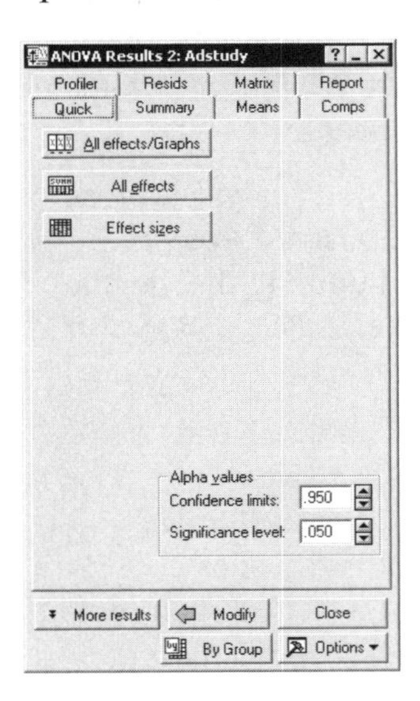

Note that this dialog contains several tabs, which enables you to quickly locate the desired results options. For example, if you want to perform planned comparisons, select the ***Comps*** tab. To view residual statistics, select the ***Resids*** tab. For this example, we will only use the results available on the ***Quick*** tab.

Reviewing ANOVA results. Let's start by looking at the ANOVA summary of all effects table by clicking the ***All effects*** button (the one with a ***SUMM***-ary icon).

Workbook1 - Repeated Measures Analysis of Variance (Adstudy)

Repeated Measures Analysis of Variance (Adstudy)
Sigma-restricted parameterization
Effective hypothesis decomposition

Effect	SS	Degr. of Freedom	MS	F	p
Intercept	3298.434	1	3298.434	497.4063	0.000000
GENDER	8.644	1	8.644	1.3035	0.259492
ADVERT	0.166	1	0.166	0.0250	0.874937
GENDER*ADVERT	0.003	1	0.003	0.0004	0.983935
Error	305.038	46	6.631		
RESPONSE	80.879	2	40.440	5.2234	0.007101
RESPONSE*GENDER	4.383	2	2.192	0.2831	0.754123
RESPONSE*ADVERT	10.286	2	5.143	0.6643	0.517097
RESPONSE*GENDER*ADVERT	8.702	2	4.351	0.5620	0.572025
Error	712.271	92	7.742		

The only effect (ignoring the ***Intercept***) in this analysis that is statistically significant (***p*** = ***.007***) is the ***RESPONSE*** effect. This result may be caused by many possible patterns of means of the ***RESPONSE*** effect (for more information, consult the *ANOVA Introductory Overview* in the *Electronic Manual*). We will now look at the marginal means for this effect graphically to see what it means.

To display the ***ANOVA Results*** dialog again (that is, "resume" the analysis), press CTRL+R, select ***Resume*** from the ***Statistics*** menu, or click the ***ANOVA Results*** button on the analysis bar. Then, click the ***All effects/Graphs*** button to display the ***Table of All Effects*** dialog to review the means for individual effects.

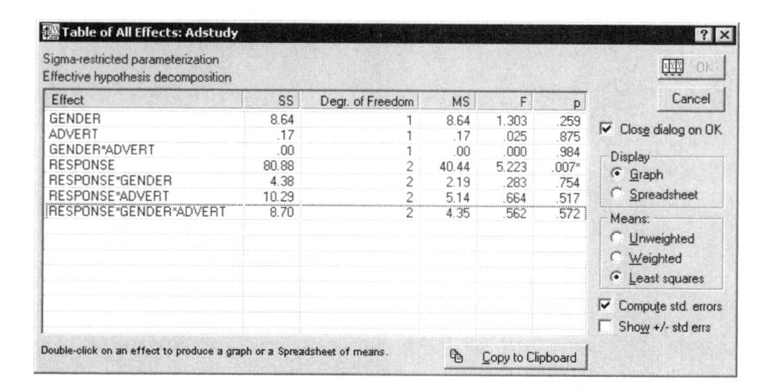

Table of All Effects: Adstudy

Sigma-restricted parameterization
Effective hypothesis decomposition

Effect	SS	Degr. of Freedom	MS	F	p
GENDER	8.64	1	8.64	1.303	.259
ADVERT	.17	1	.17	.025	.875
GENDER*ADVERT	.00	1	.00	.000	.984
RESPONSE	80.88	2	40.44	5.223	.007*
RESPONSE*GENDER	4.38	2	2.19	.283	.754
RESPONSE*ADVERT	10.29	2	5.14	.664	.517
RESPONSE*GENDER*ADVERT	8.70	2	4.35	.562	.572

This dialog contains a summary table of all effects (with most of the information you have seen in the all effects spreadsheet) and is used to review individual effects

from that table in the form of the plots of the respective means (or, optionally, spreadsheets of the respective mean values).

Plot of means for a main effect. Double-click on the significant main effect ***RESPONSE*** (the one marked with an asterisk in the ***p*** column) to see the respective plot.

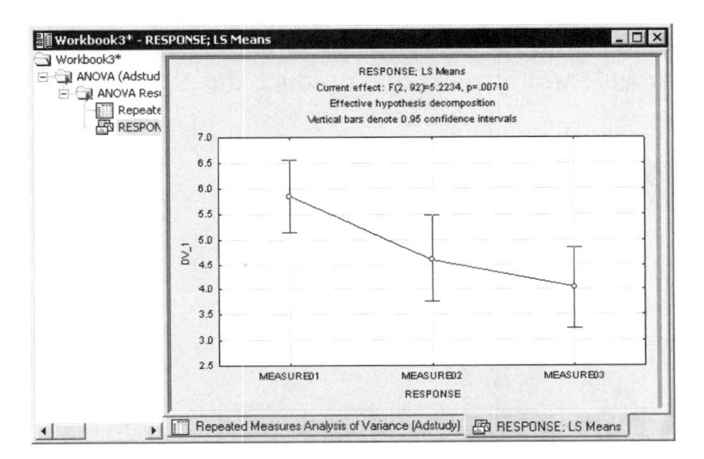

The graph indicates that there is a clear decreasing trend; the means for the consecutive three questions are gradually lower. Even though there are no significant interactions in this design (see the discussion of the ***Table of all effects***, above), we will look at the highest-order interaction to examine the consistency of this strong decreasing trend across the between-group factors.

Plot of means for a three-way interaction. To see the plot of the highest-order interaction, in the ***Table of All Effects*** dialog, double-click on the row marked *RESPONSE*GENDER*ADVERT*, representing the interaction between factors 1 (*Gender*), 2 (*Advert*), and 3 (*Response*). An intermediate dialog, ***Specify the arrangement of the factors in the plot***, is displayed, which is used to customize the default arrangement of factors in the graph (note that, unlike the previous plot of a simple factor, the current effect can be visualized in a variety of ways).

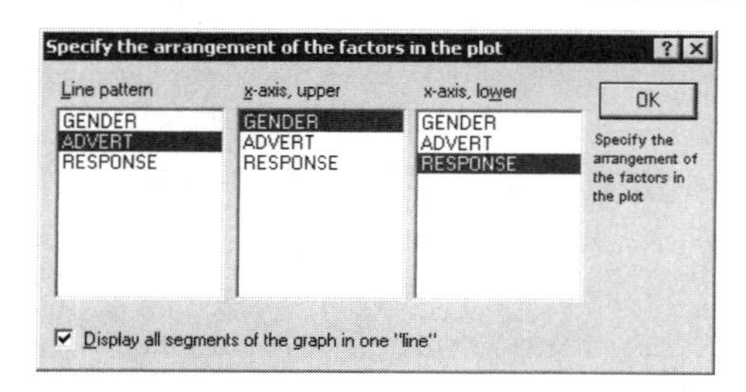

Click the ***OK*** button to accept the default arrangement and produce the plot of means.

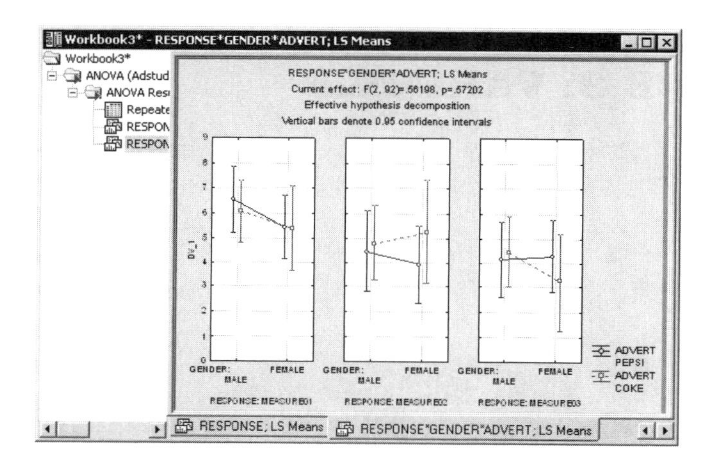

As you can see, this pattern of means (split by the levels of the between-group factors) does not indicate any salient deviations from the overall pattern revealed in the first plot (for the main effect, ***RESPONSE***). Now you can continue to interactively examine other effects – run post-hoc comparisons, planned comparisons, extended diagnostics, etc. – to further explore the results.

Interactive data analysis in *STATISTICA*. This example illustrates the way in which *STATISTICA* supports interactive data analysis. You are not forced to specify all output to be generated before seeing any results. Even simple analysis designs can produce large amounts of output and countless graphs, but usually you cannot know what will be of interest until you have a chance to review the basic output. With *STATISTICA*, you can select specific types of output, interactively conduct follow-up tests, and run supplementary "what-if" analyses after the data are processed and basic output reviewed. *STATISTICA*'s flexible computational procedures and wide selection of options used to visualize any combination of values from numerical output offer countless methods to explore your data and verify hypotheses.

Automating analyses (macros and *STATISTICA* Visual Basic). Any selections that you make in the course of the interactive data analysis (including both specifying the designs and choosing the output options) are automatically recorded in the industry standard Visual Basic code. You can save such macros for repeated use (you can also assign them to toolbar buttons, modify or edit them, combine with other programs, etc.). For more information, see Chapter 9 – *STATISTICA Visual Basic* on page 227 or the *STATISTICA Visual Basic Primer*.

EXAMPLE 3: VARIABLE BUNDLES

STATISTICA offers a unique option – variable bundles – to locate a subset of data quickly and easily in a large data file. Bundles can be created to organize large sets of variables and to facilitate the repeated selection of the same set of variables.

Open *EnginePerformance.sta*. This data set describes the performance of large engines, and contains various process parameters recorded during their manufacture. It includes 128 engines; their *Efficiency*, *Fuel Economy*, and *Power* as measured during testing; and 74 process parameters collected during the manufacture of each engine.

For this example, we will proceed with the premise that we often need to generate analyses in which the same set of variables is repeatedly used.

From the ***Data*** menu, select ***Bundle Manager*** to display the ***Variable Bundle Manager*** dialog.

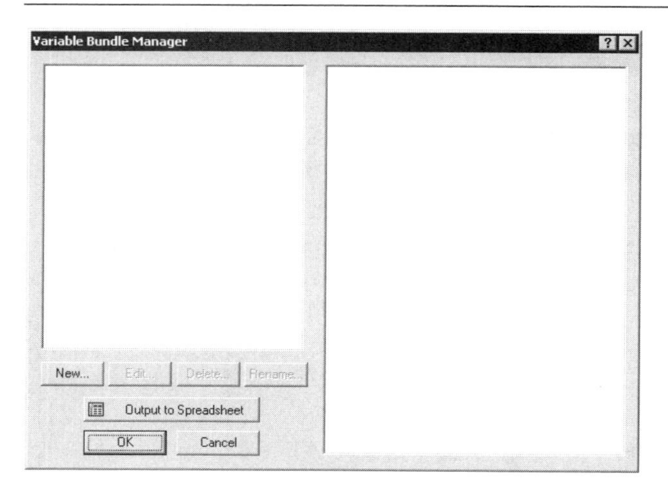

Click the ***New*** button to display the ***New Bundle*** dialog,

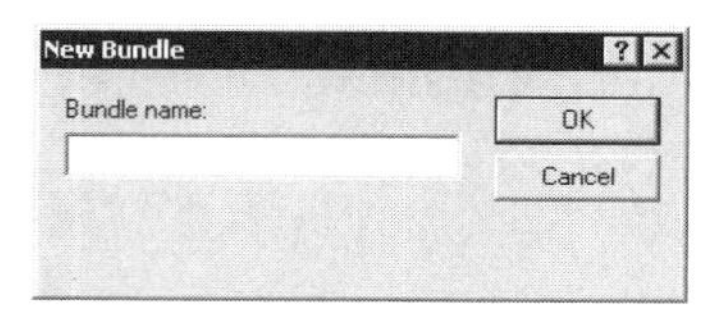

enter the name *Production* in the ***Bundle name*** field, and click the ***OK*** button. The ***Select variables for bundle*** dialog is displayed, which contains all the variables in the *EnginePerformance.sta* data set.

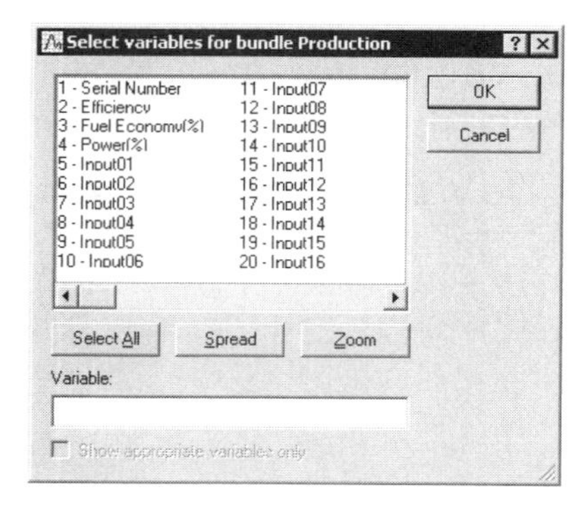

For our analyses, we need to select the variables *Input01-Input05*, *Input20*, *Input30-Input35*, and *Input70*. You can select these variables using the standard Windows SHIFT+click and CTRL+click conventions to select ranges and discontinuous lists of items, respectively.

Click the ***OK*** button to close the ***Select variables for bundle*** dialog and return to the ***Variable Bundle Manager***.

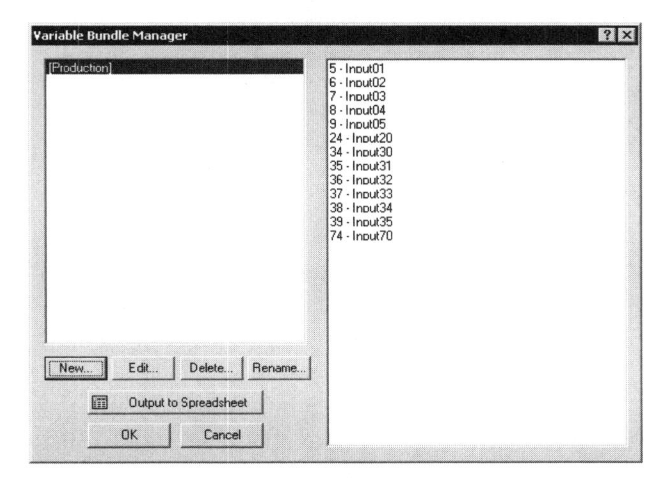

The left pane of this dialog displays the names of all bundles that have been defined for this spreadsheet (you can create numerous bundles in each spreadsheet if needed). The right pane displays the contents of the bundle that is currently selected in the left pane. If both of these panes are empty, no bundles have been created for this spreadsheet.

You can add bundles to the spreadsheet by clicking the ***New*** button, make changes to a bundle by clicking the ***Edit*** button, discard bundles by clicking the ***Delete*** button, and change the title of a bundle by clicking the ***Rename*** button. Click the ***Output to Spreadsheet*** button to create and display a spreadsheet containing information about the bundles for the active data spreadsheet.

For this example, click the ***OK*** button to accept the bundle we created and close the ***Variable Bundle Manager*** dialog. Then, from the ***Statistics*** menu, select ***Multiple Regression*** to display the ***Multiple Linear Regression*** dialog. On the ***Quick*** tab, click the ***Variables*** button to display the variable specification dialog.

Bundles are displayed in brackets and listed (in alphabetical order) at the top of the variable list. From the ***Independent variable list***, select the *Production* bundle to specify - in one click of the mouse button - Input01-Input05, Input 20, Input 30-Input35, and Input 70 as the independent variables for the analysis.

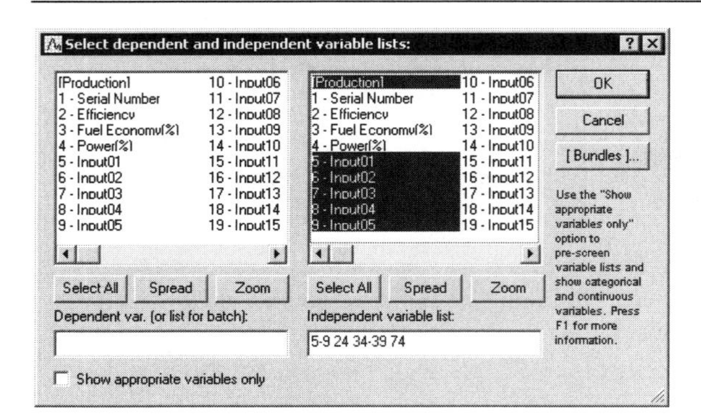

If you aren't sure what variables are included in a bundle, move the mouse pointer over the bundle name in the variable selection dialog and a ToolTip will display the variable numbers.

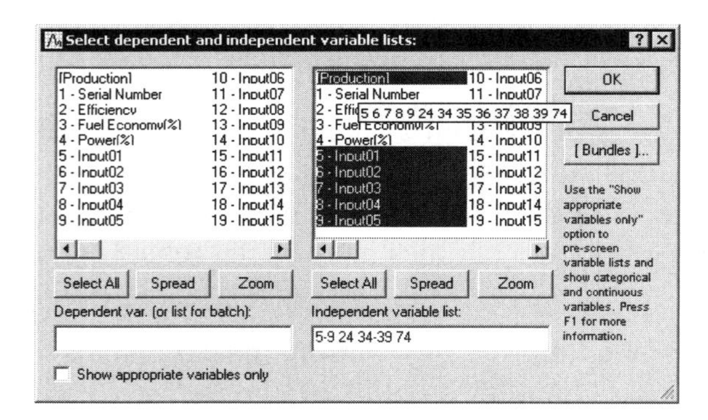

Additionally, you can view the list of variables (by name) in the ***Variable Bundles Manager*** by clicking the ***[Bundles]*** button. Note that bundles are defined for a single spreadsheet, and they are only used for variable selection. Hence, they are never listed in reports or outputs.

As you can see with this example, you will save considerable time by selecting a bundle rather than looking for the correct variables to choose in a large data set.

StatSoft®

EXAMPLE 4: BY-GROUP ANALYSES

STATISTICA offers a powerful option to turn *every* statistical or graphics analysis into an analysis by group. When reviewing results on the results dialog of practically any analysis, or using the ***Graphs*** options, you can select one or more grouping variables, and then create results 1) for all cases in the data combined, and/or 2) broken down by each combination of unique values in the grouping variables.

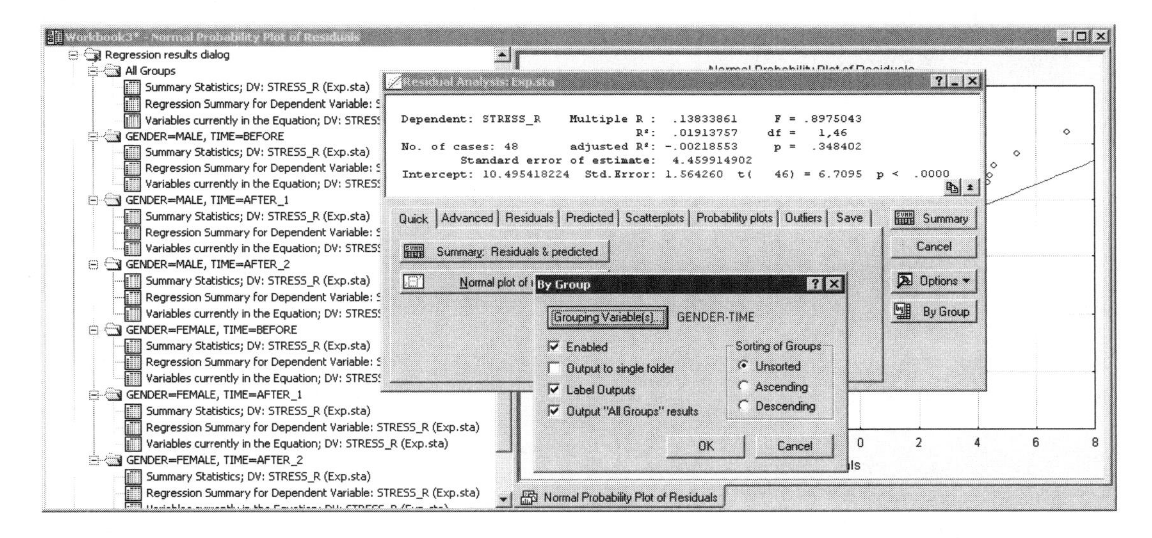

This is a very powerful new tool for interactive and exploratory data analysis, allowing you to review quickly whether any patterns or specific results hold in all subgroups, samples, or strata in your data.

For example, you may be performing a multiple regression analysis, and decide to review, without exiting the current dialog, the results broken down by *Gender* and some other grouping variable in your data. After selecting (enabling) this option (by clicking the ***By Group*** button), every time you select any of the results buttons (e.g., to create a summary results spreadsheet or graph), all results are computed not only for all groups (optionally), but also for each unique combination of grouping variables that were specified (e.g., by *Gender* and some other grouping variable).

The results of the *By Group* analysis can be placed either in the default results workbook into their own folder, labeled with the respective by-group condition (e.g., *Gender=Female; Time=After1*), or can be placed into the same folder with all other results.

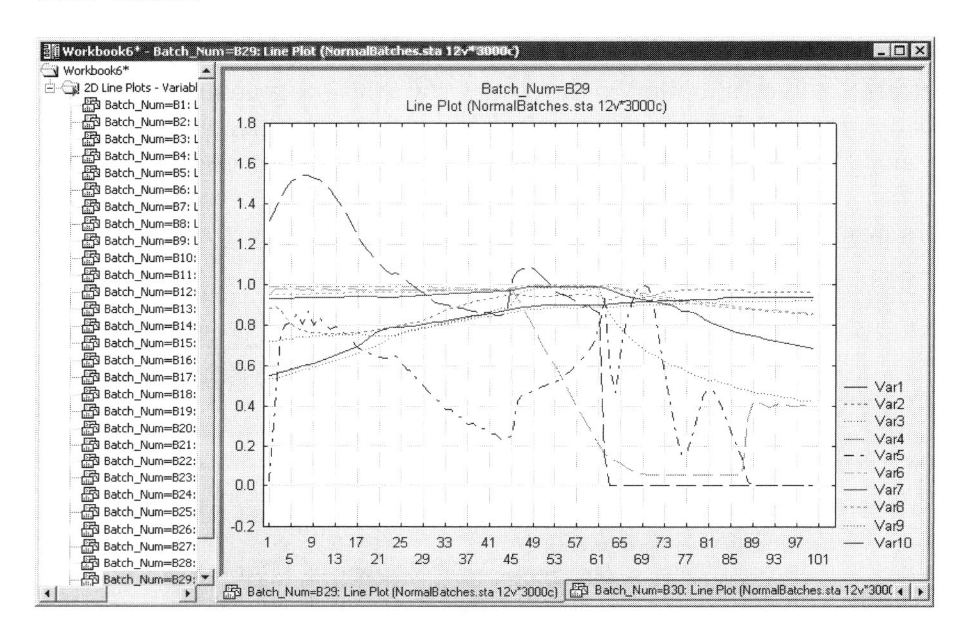

For example, you could create multiple line plots to describe a multivariate batch process, creating a separate graph ("trajectories") for each batch.

Exploring Experimental Data Using the By-Group Option

This example is based on the data file *Tomatoes.sta*, which is one of the example data files described in greater detail in the *Experimental Design* section of the *STATISTICA Electronic Manual* (see the example *Designing and Analyzing a* $2^3 3^2$ *Experiment*). Connor and Young (in McLean and Anderson, 1984) report an experiment (taken from Youden and Zimmerman, 1936) on various methods of producing tomato plant seedlings prior to transplanting in the field.

Start by opening the example *Tomatoes.sta* data set. Select ***Open Examples*** from the ***File*** menu item to display the ***Open a STATISTICA Data File*** dialog. Double-click on the *Datasets* file, and then select and open the *STATISTICA* data set *Tomatoes.sta*.

Data: Tomatoes (6v by 36c)

Tomato production as function of soil, pot size, variety, method, and location

	1 SOIL CONDITION	2 POTSIZE	3 VARIETY	4 PRODUCTION METHOD	5 LOCATION	6 POUNDS
1	Field	Three	Bonny	Flat	A	85.9
2	Field	Four	Marglobe	Flat	A	99.3
3	Plus	Three	Marglobe	Flat	A	119.8
4	Plus	Four	Bonny	Flat	A	115.5
5	Field	Three	Bonny	Fibre	C	118.3
6	Field	Four	Marglobe	Fibre	C	115.4
7	Plus	Three	Marglobe	Fibre	C	184.9
8	Plus	Four	Bonny	Fibre	C	161.7
9	Field	Three	Bonny	FibrePl	B	127.6
10	Field	Four	Marglobe	FibrePl	B	166.8
11	Plus	Three	Marglobe	FibrePl	B	158.6

Shown here are a few rows (cases) of that data file. You can refer to the *Experimental Design Electronic Help* example topic for a complete analysis of these data.

Exploring Patterns by Variety

This example illustrates a typical "workflow" as it often applies to the analysis of discrete or batch-manufacturing data, i.e., the goal of the analysis is to verify (graphically or analytically) that some patterns or distributions equally apply to all samples, parts, or batches.

We will explore the effect of *Production Method*, *Soil Condition*, and *Potsize* on yield (*Pounds*), and evaluate whether any patterns hold for each *Variety* in the study. Instead of performing a complete analysis of variance (as is described in the *Experimental Design* example of the *Electronic Help*), we will use mostly graphical methods and visual inspection.

Specifying variability plots. From the ***Graphs - 2D Graphs*** submenu, select ***Variability Plots*** to display the ***Variability Plot*** dialog. Click the ***Variables*** button, and in the ***Select Variables for Variability Plot*** dialog, select *Pounds* as the ***Dependent variable***, and *Soil Condition*, *Potsize*, and *Production Method* from the ***Grouping variable*** list for this graph.

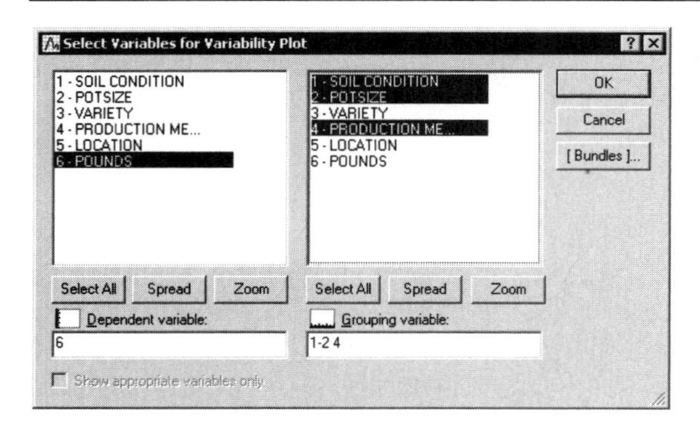

Further on in the example, we will create the graph by *Variety* to illustrate the *By Group* features. Now, click the ***OK*** button in the variable selection dialog.

Reordering variables for variability plot. For the most informative plot, let's reorder the variables so that *Production Method* will be the first factor in the list of ***Factors***. Click on that variable in the ***Factors*** list, and then, while pressing the left mouse button, drag it to the top of the list.

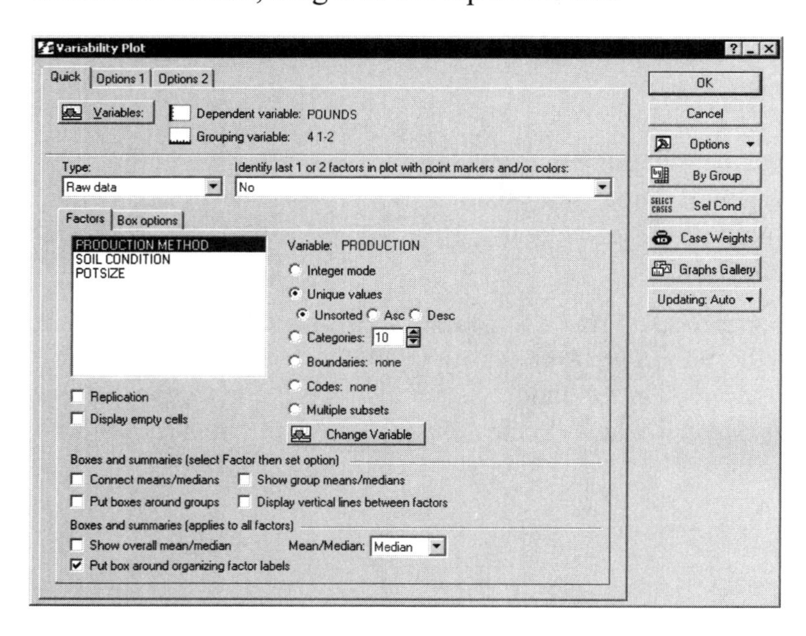

Finally, also in the ***Variability Plot*** dialog, ensure that *Production Method* is selected in the ***Factors*** list, and select the ***Put boxes around groups*** check box.

Specifying "by-grouping." We want to create the variability plot for *Production Method, Soil Condition*, and *Potsize* for all varieties of tomatoes combined, and broken down by *Variety* (one graph per *Variety*). Click the ***By Group*** button to display the ***By Group*** dialog.

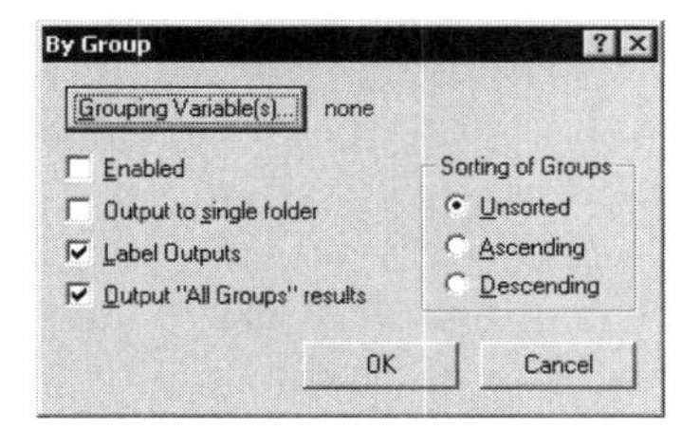

Click the ***Grouping Variable(s)*** button to display the ***Select By Variables*** dialog, and specify *Variety* as the *By Group* variable.

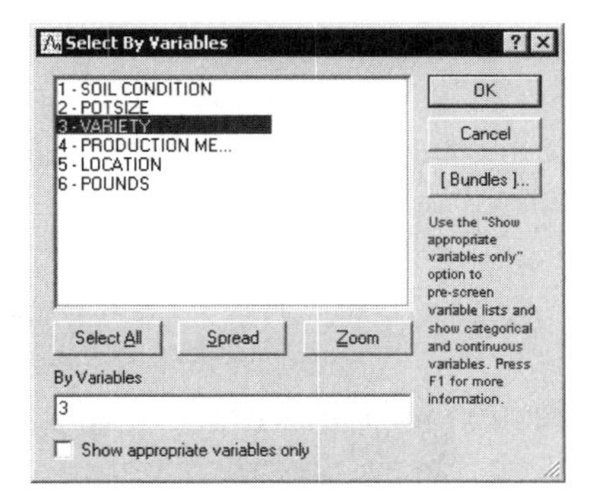

Note that you can specify more than one *By Group* variable, in which case all subsequent analysis will be performed broken down by each unique combination of values found in the *By Group* variables.

Reviewing the variability plots. Now click ***OK*** to close the ***Select By Variables*** dialog, and click ***OK*** to close the ***By Group*** dialog. In the ***Variability Plot*** dialog, click ***OK*** to create the graphs.

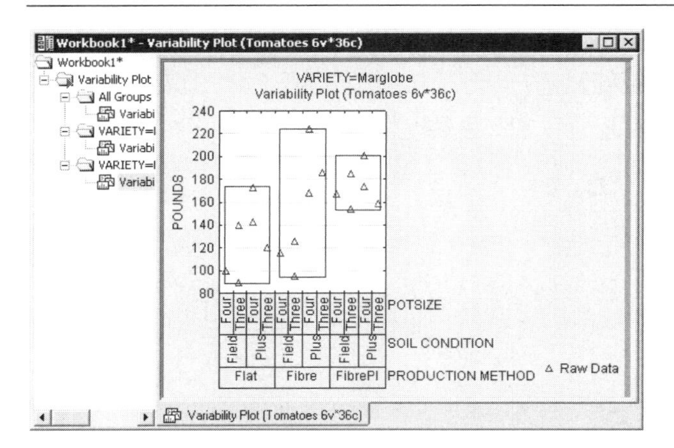

Notice how the *Variability Plot* is created 1) for *All Groups*, and 2) for each *Variety* (*Bonny*, and *Marglobe*).

If you review these graphs carefully, you will see that the *Production Method* appears to make little difference (in the observed values for *Pounds*) for *Variety=Bonny*, while for *Variety=Marglobe*, the *FibrePl* method shows the least variability in values, which are generally at the higher end of the distribution of all values for variable *Pounds*.

Descriptive Statistics By Group

Let's next use the ***Descriptive Statistics*** options to further explore this. From the ***Statistics*** menu, select ***Basic Statistics/Tables*** to display the ***Basic Statistics and Tables*** Startup Panel. Select ***Breakdown & one-way ANOVA***, and click the ***OK*** button to display the ***Statistics by Groups (Breakdown)*** dialog. Click the ***Variables*** button, and in the ***Select the dependent variables and grouping variables*** dialog, specify *Pounds* as the ***Dependent variable*** and *Production Method* as the ***Grouping variable***. Then click ***OK*** to close the variable selection dialog, and click ***OK*** in the ***Statistics by Groups (Breakdown)*** dialog to display the ***Statistics by Groups - Results*** dialog.

We want to compute *Statistics By Groups*, broken down further by tomato *Variety*. So, click the ***By Group*** button, and in the ***By Group*** dialog, click the ***Grouping Variable(s)*** button. In the ***Select By Variables*** dialog, select *Variety* as the *By Group Variable*.

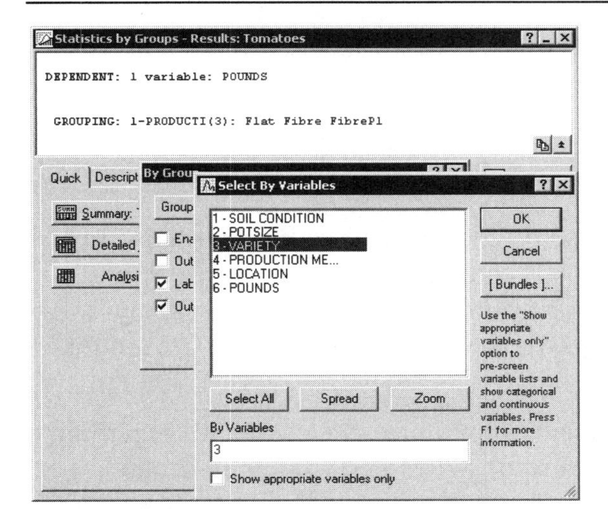

Now, click ***OK*** in this dialog and click ***OK*** in the ***By Group*** dialog. In the ***Statistics by Groups - Results*** dialog, click in sequence, 1) the ***Summary*** button, 2) the ***Analysis of Variance*** button, and 3) the ***Interaction plots*** button.

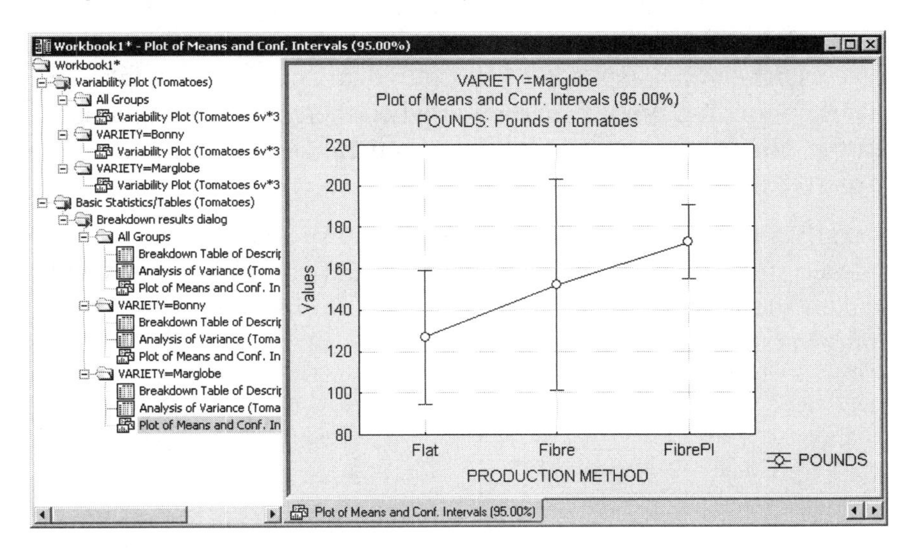

All results are placed into the respective folder, either under the *All Groups* folder, or the *Variety=Bonny* or *Variety=Marglobe* folders.

You can now review these results for all groups combined and broken down by *Variety*; as you will see, indeed, *Production Method* appears to have an effect on yield (*Pounds*) for *Variety=Marglobe*, while there is no indication of such an effect for *Variety=Bonny*.

Summary

With *STATISTICA*, you can perform ad-hoc by-group analyses from virtually any results dialog, reviewing results for all groups combined or broken down by one or more grouping variable. This very powerful feature for exploratory data analysis can be used to compare groups and verify consistency of results across groups for any analysis.

Before concluding this topic, a few comments about the technical details regarding the implementation of this feature may be useful. When performing by-group analyses, as illustrated in this example, the program will actually rerun the analyses for each group (and all groups), leveraging the *STATISTICA* Visual Basic macro code that is recorded automatically during the interactive analyses, and which can be saved as macros as described elsewhere in this manual (see Chapter 9 - *STATISTICA Visual Basic*). When analyzing very large data problems (e.g., very large unbalanced experimental designs or complex analyses that require iterated computations before results can be displayed), the individual analyses may take up significant amounts of computing time, in particular when there are many unique groups identified in the data (e.g., imagine a complex generalized linear model estimated for each of 100 groups).

Therefore, it is generally a good idea to begin each exploratory analysis by computing simple descriptive statistics, frequency tables, and graphs to understand the structure of the data and identify the number of unique groups (combination of values in the grouping variables) in the data.

3 CHAPTER

STEP-BY-STEP EXAMPLES: ADVANCED

CHAPTER

STEP-BY-STEP EXAMPLES: ADVANCED

EXAMPLE 1: INPUT DATA DIRECTLY FROM EXCEL

In addition to using the traditional *STATISTICA* spreadsheet, you can open Excel files in a *STATISTICA* window and then perform analyses using the Excel file as your data source.

From the *STATISTICA* ***File*** menu, select ***Open Examples*** to display the ***Open a STATISTICA Data File*** dialog. From the ***Files of type*** drop-down list at the bottom of the dialog, select *Excel Files (*.xls)*. Double-click the *Datasets* folder, and then select the *Weather report* data file, which is an Excel file. Click the ***Open*** button, and the ***Opening file*** dialog will be displayed. Click the ***Open as an Excel Workbook*** button, and the Excel file will be displayed. Note that when an Excel worksheet is opened in *STATISTICA*, the Excel and *STATISTICA* menus merge, enabling you to access key functionality for both applications.

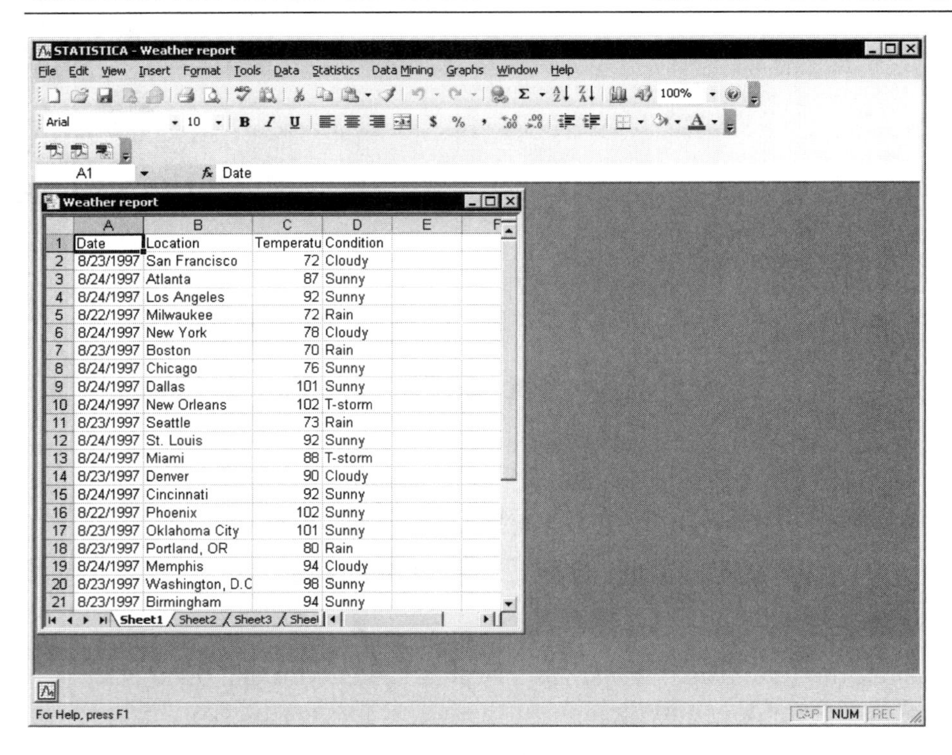

	A	B	C	D
1	Date	Location	Temperatu	Condition
2	8/23/1997	San Francisco	72	Cloudy
3	8/24/1997	Atlanta	87	Sunny
4	8/24/1997	Los Angeles	92	Sunny
5	8/22/1997	Milwaukee	72	Rain
6	8/24/1997	New York	78	Cloudy
7	8/23/1997	Boston	70	Rain
8	8/24/1997	Chicago	76	Sunny
9	8/24/1997	Dallas	101	Sunny
10	8/24/1997	New Orleans	102	T-storm
11	8/23/1997	Seattle	73	Rain
12	8/24/1997	St. Louis	92	Sunny
13	8/24/1997	Miami	88	T-storm
14	8/23/1997	Denver	90	Cloudy
15	8/24/1997	Cincinnati	92	Sunny
16	8/22/1997	Phoenix	102	Sunny
17	8/23/1997	Oklahoma City	101	Sunny
18	8/23/1997	Portland, OR	80	Rain
19	8/24/1997	Memphis	94	Cloudy
20	8/23/1997	Washington, D.C	98	Sunny
21	8/23/1997	Birmingham	94	Sunny

From the ***Statistics*** menu, select ***Basic Statistics/Tables***. The ***Excel file*** dialog will be displayed.

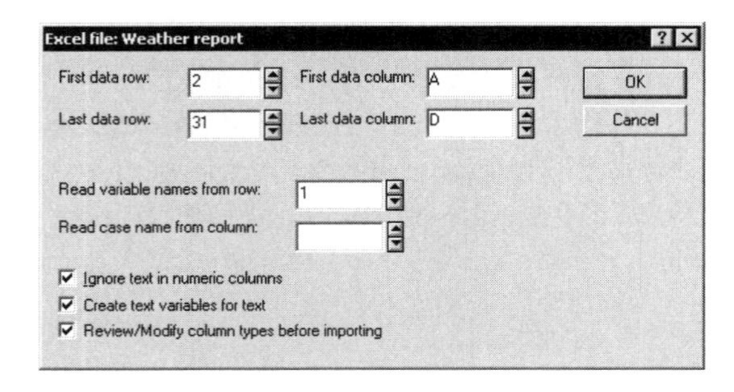

This dialog is displayed whenever you select an option from the ***Statistics***, ***Data-Mining***, or ***Graphs*** menu after opening an Excel worksheet in the *STATISTICA* window. Note that *STATISTICA* has determined the logical specifications, but these options can be changed if necessary. When variable names are not included with the Excel worksheet, *STATISTICA* will assign variable names, i.e., *Var1*, *Var2*, *Var3*, etc. As with *STATISTICA* spreadsheets, all values in a column will be used for the

selected analysis unless case selection conditions are specified. For this example, click the ***OK*** button in the ***Excel file*** dialog to accept the defaults; the dialog will close, and the ***Review/Edit Column Types*** dialog will be displayed.

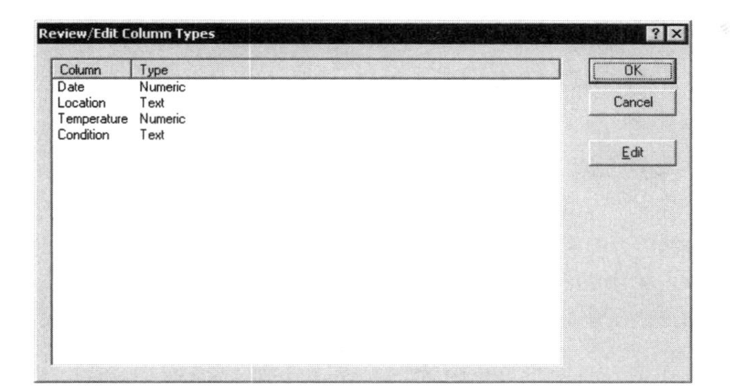

STATISTICA allows you to define the data type for the specific columns. Data types include numeric, text, mixed numeric and text, and missing data. Empty cells in an Excel worksheet are always treated as missing data, and when a numeric column contains text values, those values are also treated as missing data. *STATISTICA* provides default data types for all columns based on the first few rows of data (in fact, you can clear the ***Review/Modify column types before importing*** check box in the ***Excel file*** dialog before clicking ***OK*** in that dialog, and the ***Review/Edit Column Types*** dialog will not be displayed). However, you can change the default types if needed: select the name of the column you want to change and click the ***Edit*** button (or double-click on the name of the column you want to change) to display the ***Change Import Column Type*** dialog, where you can specify the type you prefer.

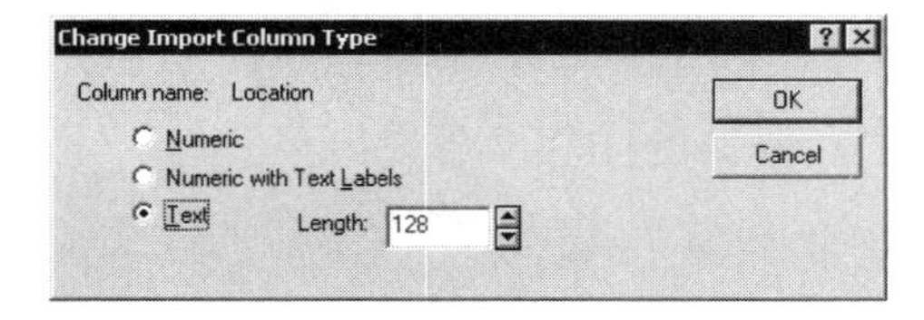

For this example we will accept the defaults, so click the ***Cancel*** button in the ***Change Import Column Type*** dialog, and click the ***OK*** button in the ***Review/Edit Column Types*** dialog. After you click ***OK***, the Startup Panel for the selected analysis or graph will be displayed (in this example, the ***Basic Statistics and Tables*** Startup Panel), and you can proceed with the analysis as usual.

EXAMPLE 2: ACCESSING DATA DIRECTLY FROM DATABASES

STATISTICA provides access to virtually all databases (including many large system databases such as Oracle, Sybase, etc.) via *STATISTICA* Query, accessible from either the ***File - Get External Data*** submenu or the ***Data - Get External Data*** submenu. For importing data from a database directly into a *STATISTICA* Spreadsheet so that it can be saved, the tool to use is *STATISTICA* Query.

With *STATISTICA* Query, you can easily access data using OLE DB conventions. OLE DB is a database architecture [based on the Component Object Model (COM)] that provides universal data integration over an enterprise's network, from mainframe to desktop, regardless of the data type.

STATISTICA Query supports multiple database tables; specific records (rows of tables) can be selected by entering SQL statements. *STATISTICA* Query automatically builds the SQL statement for you as you select the components of the query via a simple graphical interface and/or intuitive menu options and dialogs. Hence, an extensive knowledge of SQL is not necessary in order for you to create advanced and powerful queries of data in a quick and straightforward manner. Multiple queries based on one or many different databases can also be created to return data to an individual spreadsheet; hence, you can maintain connections to multiple external databases simultaneously.

For this example, create a new database query: from the ***File - Get External Data*** submenu, select ***Create Query***. *STATISTICA* Query will start, and the ***Database Connection*** dialog will be displayed.

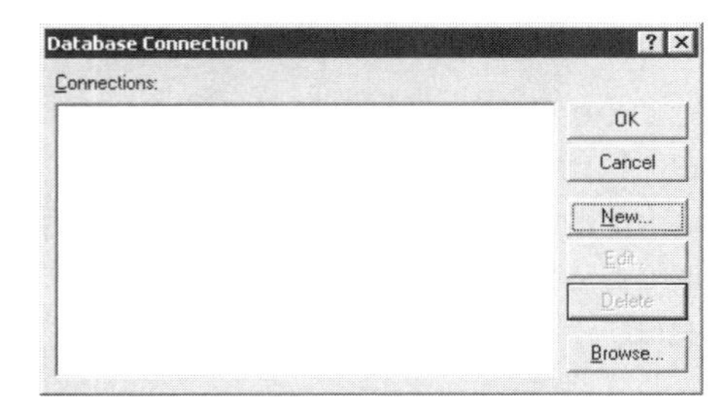

From this dialog, you can choose existing database connections or define new ones. For this example, we'll create a new database connection, so click the ***New*** button to display the ***Data Link Properties*** dialog.

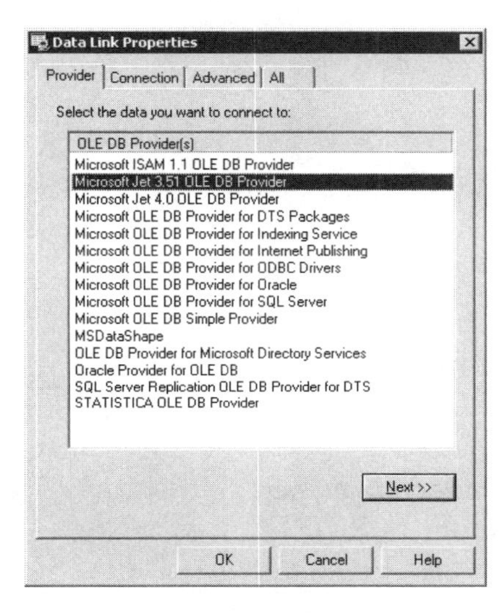

You can choose either the OLE DB provider that was supplied by your database vendor, or one of the Microsoft default OLE DB providers that is compatible with your database system. For this example, we'll use a Jet database sample file installed with *STATISTICA*, so select ***Microsoft Jet 3.51 OLE DB Provider*** and click the ***Next >>*** button. The ***Data Link Properties*** dialog - ***Connection*** tab will be displayed.

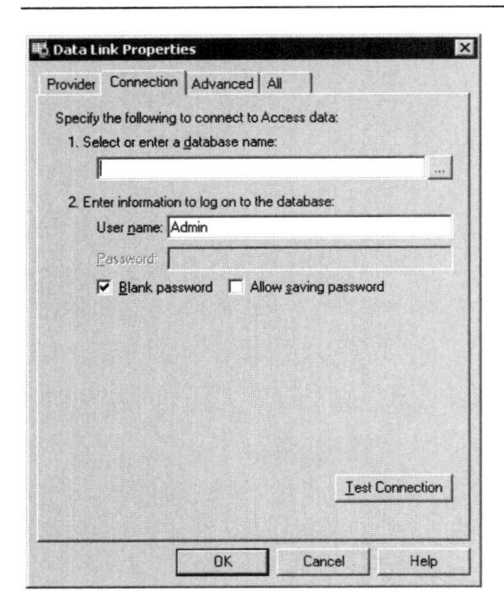

Click the [...] button adjacent to the ***Select or enter a database name*** edit box to display the ***Select Access Database*** dialog. Browse to the *STATISTICA Examples* folder, open the *Database* folder, and double-click on the *ProcessData.mdb* file. The path and file name will be displayed in the ***Select or enter a database name*** edit box.

Click ***OK*** on the ***Data Link Properties*** dialog to display the ***Add a Database Connection*** dialog. Enter ***Process Name*** in the ***Name*** edit box, and click ***OK***. The ***Database Connection*** dialog will be displayed again, with the new *Process Data* connection defined. Select this connection, and click ***OK***. The *STATISTICA* Query window will be displayed, with all the database tables in the tree view on the left:

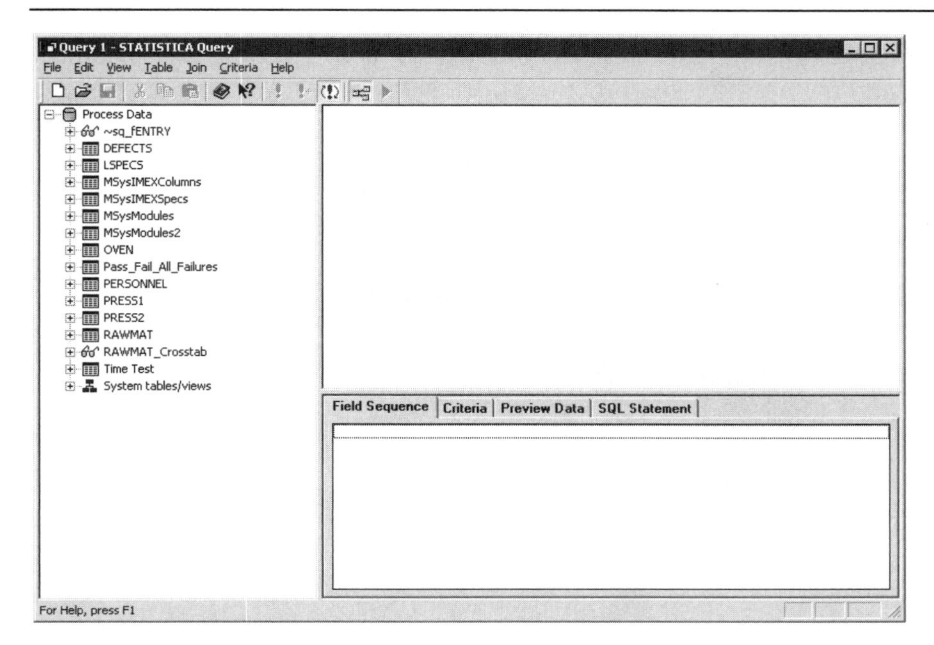

Right-click on the *RAWMAT* table, and from the shortcut menu, select ***ADD*** to add the table to the table view pane (the upper-right pane in the *STATISTICA* Query window). Then, right-click on the *PERSONNEL* table, and add it to the table view pane.

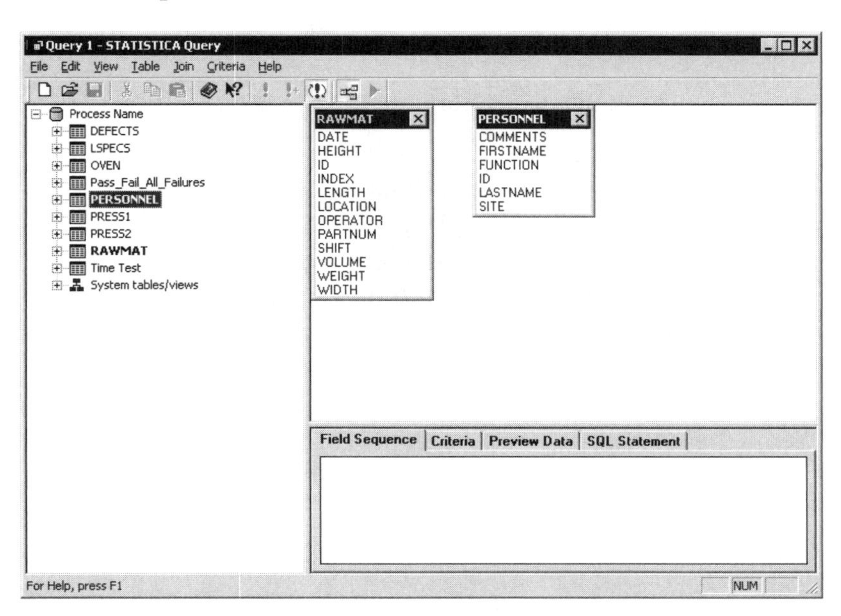

To select the fields to include in the query, right click on the *RAWMAT* table in the table view pane, and from the shortcut menu, select ***Select All Fields***; then do the same for the *PERSONNEL* table. Note that both tables contain an *ID* field by which they can be joined. To join them, select *ID* in the left *RAWMAT* table, and drag it to the *ID* field in the *PERSONNEL* table. Click the ***Preview Data*** tab in the lower-right pane to view a preview of the query data:

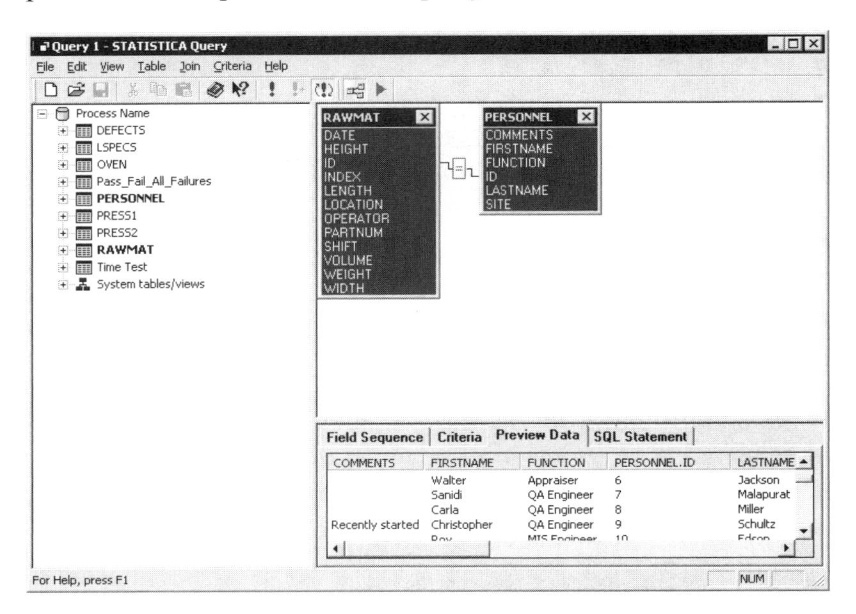

Click the ***SQL Statement*** tab to display the SQL Statement generated by the query.

To return the data to a *STATISTICA* Spreadsheet, click the green arrow on the *STATISTICA* Query toolbar. The ***Returning External Data to Spreadsheet*** dialog will be displayed, where you can control whether the query will be placed into a new or current spreadsheet, and can adjust other query parameters. Select the ***New Spreadsheet*** option button, and click the ***Run Now*** button to run the query. After a few moments, the data is returned to the *STATISTICA* Spreadsheet.

Data: Query 1 (18v by 29c)

	2 FIRSTNAME	3 FUNCTION	4 PERSONNEL.ID	5 LASTNAME	6 SITE	7 RAWMAT.ID	8 DATE	9 HEIGHT	10 INDEX	11 LENGT
1	Peter	QA Engineer	1	Barry	Boston	1	2/6/1999	9.14793329	0.569596665	239.589
2	Lynn	MIS Enginee	4	Hawks	Dallas	4	2/9/1999	10.4420302	0.6088003	231.170
3	Frank	Appraiser	5	Peters	Dallas	5	2/10/1999	10.3950985	0.5810556	225.492
4	Walter	Appraiser	6	Jackson	Dallas	6	2/11/1999	10.264723	0.528721226	224.758
5	Sanidi	QA Engineer	7	Malapurat	Dallas	7	2/12/1999	10.2599724	0.577218758	224.881
6	Carla	QA Engineer	8	Miller	Dallas	8	2/13/1999	11.3507885	0.67666181	223.772
7	Christopher	QA Engineer	9	Schultz	Dallas	9	2/14/1999	7.62945273	0.447120064	228.864
8	Roy	MIS Enginee	10	Edson	Dallas	10	2/15/1999	9.58322023	0.55163456	229.451
9	Gabrielle	MIS Enginee	11	Smith	Dallas	11	2/16/1999	11.287434	0.685260751	234.274
10	Frank	QA Engineer	12	Foster	Dallas	12	2/17/1999	10.8893952	0.626008716	234.566
11	Herb	QA Engineer	13	Ellis	Dallas	13	2/18/1999	7.88062345	0.428575782	234.195
12	John	Operator	14	Klein	Dallas	14	2/19/1999	11.1052502	0.621735308	232.982
13	Alva	Operator	15	Fernandez	Dallas	15	2/20/1999	10.2465437	0.573303637	226.972
14	John	Operator	16	Kloster	Boston	16	2/21/1999	8.70109242	0.473787209	229.255
15	Yang	Operator	17	Chen	Boston	17	2/22/1999	9.82721357	0.534349322	228.257
16	Angie	Operator	18	Dawson	Boston	18	2/23/1999	10.4395011	0.618907147	231.544
17	Oscar	Operator	19	Stevens	Dallas	19	2/24/1999	10.0111308	0.586096374	226.122
18	Adolfo	Operator	20	DiLorenzio	Dallas	20	2/25/1999	8.06192367	0.451578571	224.22
19	Manuel	Operator	21	Ricardo	Dallas	21	2/26/1999	11.0710936	0.634628312	226.078
20	Frank	Operator	22	Schwimmer	Dallas	22	2/27/1999	9.32197665	0.525140435	225.281

Now the data can be analyzed with any of the *STATISTICA* tools. Note that the spreadsheet retains the database connection, and you can re-run the query at any time by selecting ***Refresh Data*** from the ***File - Get External Data*** submenu or by pressing F5 on your keyboard when the spreadsheet is open.

EXAMPLE 3: DATA PREPARATION – CLEANING AND FILTERING

Summary of Options for Data Filtering/Recoding

In practice, most of the time required to complete a data analysis or data mining project is spent on the preparation of data. Sometimes as much as 90% of all time and effort required to complete a project is related to the proper cleaning and preparation of the data.

When building prediction models using data mining tools, or even when just computing simple descriptive statistics (averages, frequency distributions), results of analyses can be very misleading if, for example, large numbers of duplicate records are included (e.g., the same part numbers are recorded multiple times), the data include outliers or miscoded values (outside the valid data ranges), or excessive numbers of missing (blank) data.

In the ***Data - Data Filtering/Recoding*** submenu, *STATISTICA* provides commands to address such data quality issues quickly and effectively so that meaningful and valid data analyses or data mining projects can be completed in less time.

Filter Duplicate Cases

Use this option when you suspect that your data file may contain duplicate records (e.g., duplicate/identical customer records).

For example, suppose that in an analysis of customer records, to identify typical customer demographics ("profiles"), you want to count each customer only once; however, your customer database is organized by transactions, so each customer may appear multiple times. In this case, you can use the *Filter Duplicate Cases* options to create a data file for the analyses, where each record is unique (e.g., where each customer ID is unique, and appears only once).

Duplicate information example. Open the *Duplicates.sta* data file. From the ***Data - Data Filtering/Recoding*** submenu, select ***Filter Duplicate Cases*** to display the ***Filter Duplicate Cases*** dialog. In the ***Input*** group box, the ***Variables*** option is used to specify the basis of distinction for duplicates; e.g., click the ***Variables*** button, and in the variable selection dialog, select *Respondent* so that all respondents will be checked for duplicates. Click ***OK*** in the variable selection dialog to return to the ***Filter Duplicate Cases*** dialog.

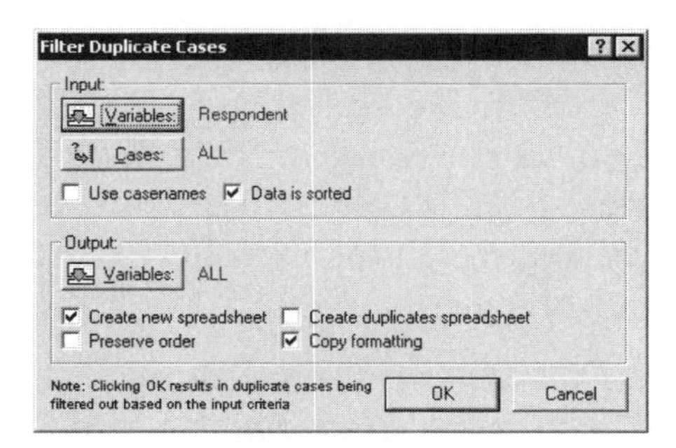

In the ***Input*** group box, click the ***Cases*** button to display the ***Spreadsheet Case Selection Conditions*** dialog, which contains options to select only specified observations or cases for the de-duping operations. In this example, we will filter all the cases, so click the ***Cancel*** button on the ***Spreadsheet Case Selection Conditions*** dialog.

The ***Use casenames*** check box is cleared by default; we will leave this option as is for this example. When this check box is selected, case names are used as one of the bases for distinction, i.e., *STATISTICA* will treat as duplicates any cases that have the same case name (provided the cases match on any other specified variables as well). When the check box is cleared, duplicate case names will be ignored.

Clear the ***Data is sorted*** check box (because the current data file has not been sorted – when you have an extremely large data file, it is more efficient to sort the data first).

In the ***Output*** group box, verify that all variables are selected (*ALL* will be adjacent to the ***Variables*** button). This option is used to select the variables in the input spreadsheet that will be included in the output (filtered) spreadsheet; the default is *ALL*.

Verify that the ***Create new spreadsheet*** check box is selected (the default), and select the ***Create duplicates spreadsheet*** check box. Leave the last two options at their defaults: the ***Preserve order*** check box is cleared [the new spreadsheets will be sorted by the variable(s) that were selected as the basis of distinction, in this example, *Respondent*], and the ***Copy formatting*** check box is selected. Click ***OK***.

Two new spreadsheets will be generated. One of the spreadsheets is 10v by 51c (10 variables by 51 cases) and contains the respondents from the original spreadsheet excluding the duplications. The other spreadsheet is 10v by 9c and contains the duplicate respondents that were extracted from the original spreadsheet.

Look at the original spreadsheet, *Duplicates.sta*, and notice that some of the variable headers – *Respondent*, *State*, and *Colors* – are formatted differently. Then look at the two new spreadsheets; the variable headers for *Respondent*, *State*, and *Colors* have the same formatting in all three spreadsheets. *STATISTICA* uses sub-setting to create the new spreadsheets and ensures that variable properties of the parent spreadsheet are maintained in the child spreadsheets.

Now, close the two new spreadsheets, but leave the *Duplicates.sta* spreadsheet open. Notice that it is 10v by 60c. From the ***Data - Data Filtering/Recoding*** submenu, select ***Filter Duplicate Cases*** to display the ***Filter Duplicate Cases*** dialog again. In the ***Input*** group box, click the ***Variables*** button, and in the variable selection dialog, select *Respondent* and click ***OK***. In the ***Input*** group box, clear the ***Data is sorted*** check box. In the ***Output*** group box, clear the ***Create new spreadsheet*** check box. Click ***OK***. The dialog closes and, instead of creating a new spreadsheet with the duplicates excluded, the *Duplicates.sta* spreadsheet is modified. All duplicate cases are removed from it; it now has 10v by 51c.

Note that the filter duplicate cases functionality does not use case sensitivity (upper case, lower case letters) for a comparison of uniqueness, i.e., if you have two respondents – C. Barrett and C. BARRETT – the second respondent will be excluded.

Filter Sparse Data

It is not uncommon that some variables (parameters, or data fields) available for (for example) predictive modeling have very few valid data. For example, in a customer database self-reported (by customers) *Income* may be recorded; however, very few customers actually volunteered their current incomes, so most of the data (in that field of the database) is blank (or missing). In manufacturing data, a data field may exist to record a specific parameter, but the sensor might be faulty for an extended period of time, recording mostly missing (invalid) data.

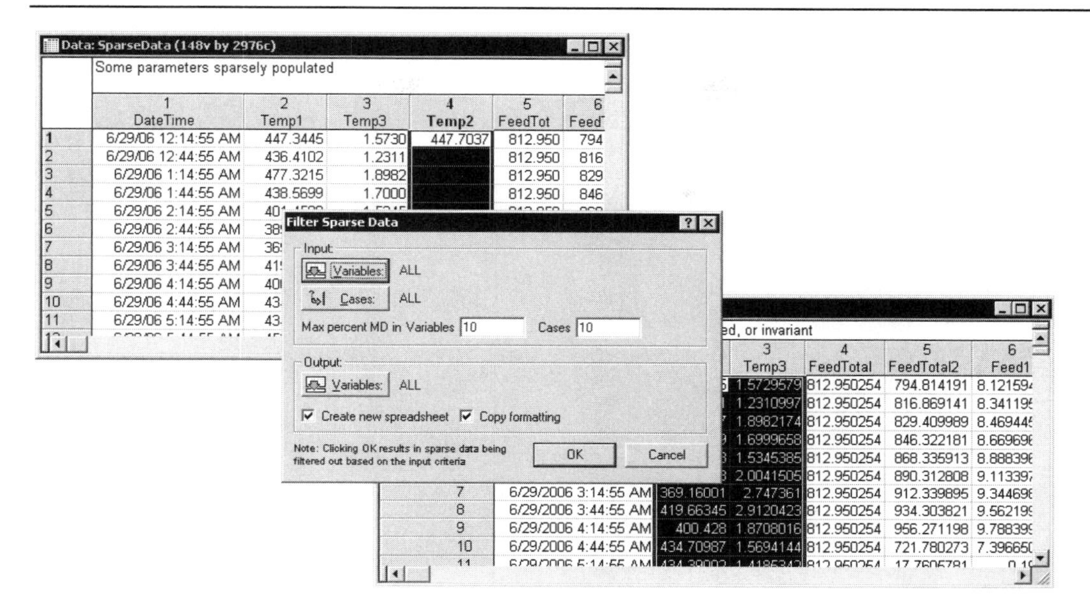

Including such "sparsely populated" (with data) variables in an analysis may lead to erroneous results, or prevent you from building predictive models altogether (depending on how the missing data are handled later in the analyses). Therefore, you may want to identify such sparse variables ahead of time using the ***Filter Sparse Data*** options (accessible from the ***Data - Data Filtering/Recoding*** submenu), and eliminate them from subsequent consideration.

Process Invariant Variables

A similar (to the sparse-data case) data quality issue that often occurs, in particular in industrial manufacturing (process) data, is that some variables (parameters) that are recorded and included in the analyses are invariant, i.e., all values are the same.

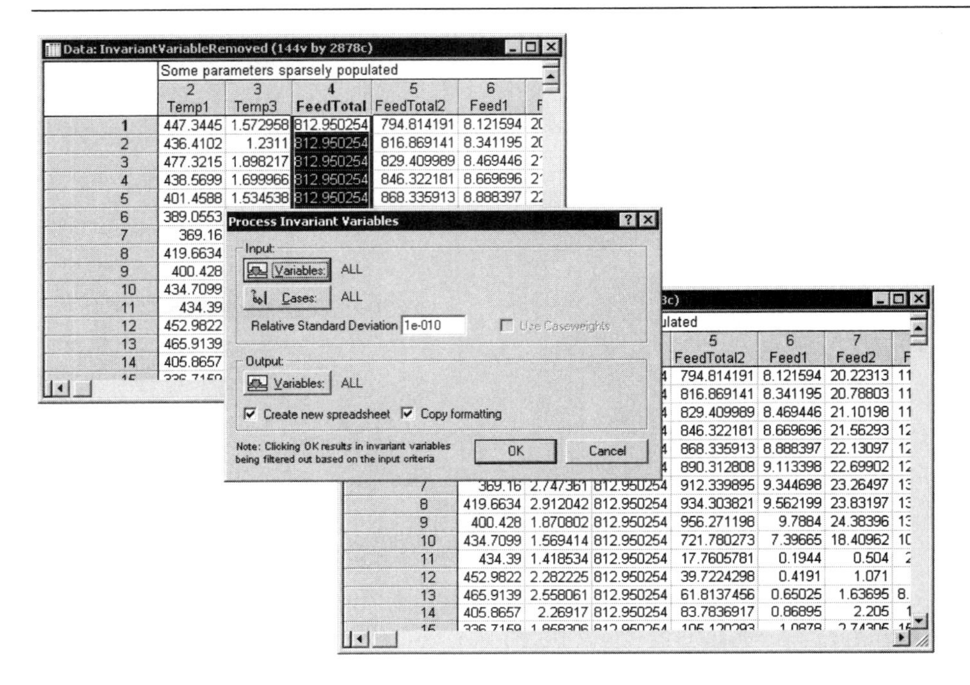

Such variables are not useful for predictive modeling, and the ***Process Invariant Variables*** options (accessible from the ***Data - Data Filtering/Recoding*** submenu) enable you to identify those variables automatically, and exclude them from further analyses.

Recode Outliers (or "Fliers")

Extreme data values or outliers (also sometimes referred to as *fliers*) can greatly affect various analyses and cause poor accuracy of prediction (data mining) models. There is no formal definition of what constitutes an "outlier" or "extreme value," and *STATISTICA*'s graphical tools may provide the best way to review data to identify such unusual observations (e.g., you could create box plots of the key variables to identify extreme observations and brush or flag them in the data).

To automatically process lists of variables to identify and remove outliers, the ***Recode Outliers*** options (accessible from the ***Data - Data Filtering/Recoding*** submenu) provide several tests for outliers (approaches for identifying extreme values).

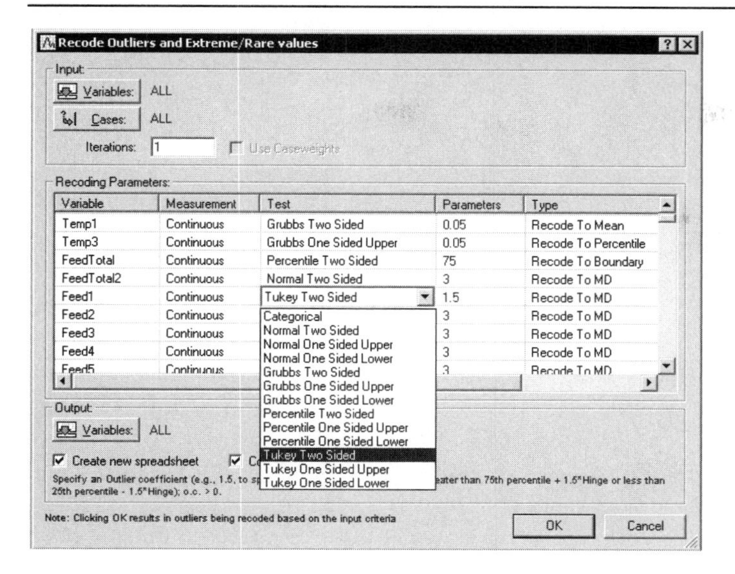

Outliers can be recoded to missing data or to valid data values (e.g., to the respective percentile boundary values, etc.).

Process Missing Data

Missing data or invalid data values must obviously be dealt with in a manner that is consistent with the goals of the analyses. In some cases, missing or invalid data may themselves provide useful information about a process or variable of interest. For example, in marketing research, it is common that respondents will refuse to provide detailed personal information regarding their health, financial dealings (e.g., savings), etc., and such refusal itself may be correlated with other significant variables of interest (e.g., refusal to answer questions related to income may itself be a good indicator of high-income, if indeed wealthier individuals in the survey tended not to answer those questions).

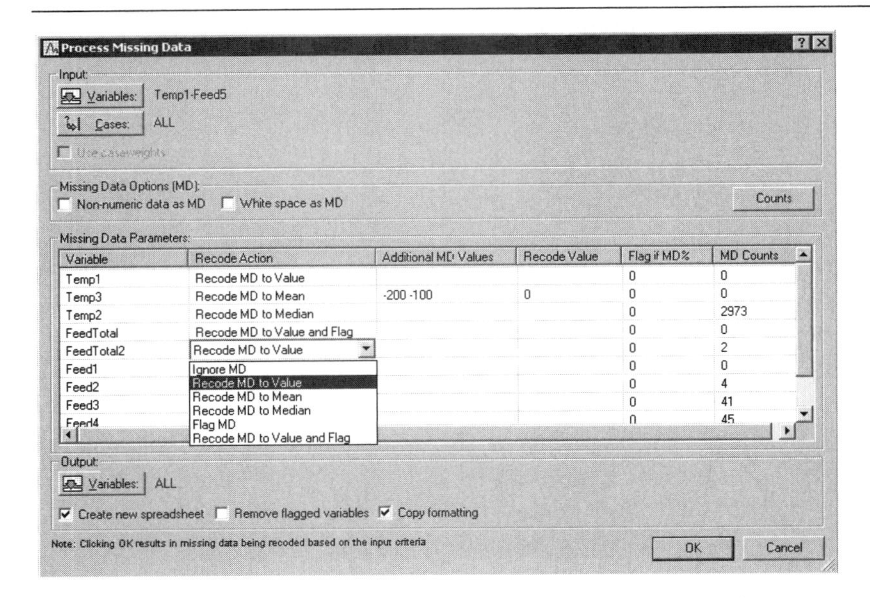

The ***Process Missing Data*** options (accessible from the ***Data - Data Filtering/Recoding*** submenu) enable you to flexibly recode missing data, define multiple missing data values or codes for a single variable (those values can then be recoded to the variable missing data code), or to just flag variables that have more than a certain percentage of missing data.

Imputation of Missing Data (*k*-Nearest Neighbor)

It is often not clear how best to recode missing data, and in fact, sometimes by recoding missing data for a particular variable to a specific value (e.g., the mean), the final results may be biased. For example, suppose in a survey all respondents who refuse to report their income tend to be in the higher income bracket. In that case, assigning the mean-income to those individuals (i.e., recoding missing data for variable *Income* to the mean income for the whole sample) may yield highly misleading results.

STATISTICA includes a very efficient method (applicable to very large data sets and databases) for replacing missing data with valid data values that are consistent with the other observations in the sample. Details regarding the *k-nearest neighbor* method and algorithm are provided in the *Electronic Help* for the *Machine Learning* module of *STATISTICA Data Miner*.

In short, using the ***MD Imputation*** options (accessible from the ***Data - Data Filtering/Recoding*** submenu), in a first pass through the data, the *k-nearest neighbor* algorithm will select a (smaller) sample from all available data. In the second pass through the data file, when missing data are encountered, they are replaced with valid (observed) values found in similar observations in the smaller sample (with respect to all other variables that were selected). So to continue this example, if indeed higher-income respondents are less likely to report this fact, but *do* report other indicators of high-income (e.g., ownership of a luxury car, more square footage of their home, etc.), then the *k-nearest neighbor* algorithm will accurately assign those individuals (who failed to report their income) to the high-income bracket.

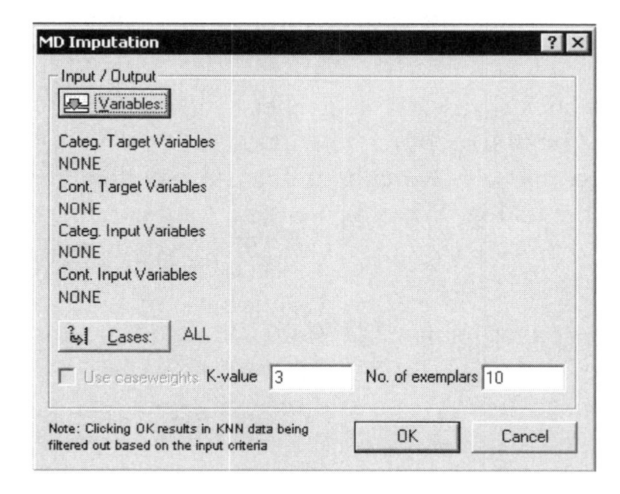

The *k-nearest neighbor* algorithm is fast and efficient, and provides an effective method for replacing missing data in the input file with "reasonable guesses" based on similar data points in the sample. This approach does not make any particular assumptions about the nature of the relationships between variables (i.e., require that a "model" be estimated for each variable, to predict missing data values), but simply uses the observed data as the model.

EXAMPLE 4: SPREADSHEET FORMULAS AND BATCH FORMULAS

You can define new variables for *STATISTICA* Spreadsheets in terms of other variables, sometimes referred to as variable transformations. Additionally you can verify data, transform data, and recode data on a single variable (as opposed to a set of transformation formulas, i.e. Batch formulas). This is done through Spreadsheet Formulas.

To access Spreadsheet Formulas, double click on a variable header in a *STATISTICA* Spreadsheet to display the ***Variable*** specification dialog. The formula is entered into the ***Long name (label or formula with Functions)*** field (also called the formula editor) located at the bottom of the dialog. When you enter a long variable name in the formula editor that starts with an equal sign, *STATISTICA* recognizes it as a formula and will verify it for formal correctness.

The formula can reference other variables either by name (*MEASURE01*, *TIME*), or by absolute variable number using the *Vx* syntax, where *x* is the absolute variable number. For example, *V3* is variable number 3. *V0* has special meaning, and refers to the current case number.

Spreadsheet Formulas are evaluated a case (row) at a time. For each case in the spreadsheet, the formula is evaluated, and references to the other variables are substituted with their values from the current case.

In *STATISTICA*, random access spreadsheet functions enable the formula to access variable values from other cases. A common example of this is the Lag function, which will reference a variable, and lag it forward or backward a certain number of cases.

The following table lists several Spreadsheet Formulas and their results.

Formula	Result
=contains(v1, "B12C")	Returns 1 if the text "B12C" is found in variable 1. Returns 0 if no match is found.
=(v1+v2+v3)/3	Compute the mean of the first three variables.
=(v0<=10)*1+(v0>10)*2	Recodes cases 1-10 as 1. The other cases are set to 2.
=((v1=1) AND (v2=5))*5	Returns the value of 5 if v1=1 and v2=5, otherwise set to 0.
=student(v4,15)	Returns probability density values of the Student's t distribution based on the values of v4 and 15 degrees of freedom.
=iif(V0 <= 1, V3, V3+LAG(VCUR, 1))	Performs a cumulative sum of variable 3.

Note that you can click the Functions button in the ***Variable*** specification dialog in order to open the ***Function Browser*** dialog, and display the complete list of Formulas and Operators (=, +, >, and, or…).

Example: Spreadsheet Formula

Open the *Adstudy.sta* data file. We will create a new variable that is the mean of variables *3* through *25* (i.e., *MEASURE01* through *MEASURE23*).

Double-click on the first blank variable header (right after variable 25). The ***Add Cases and/or Variables*** dialog will be displayed. Click the ***OK*** button to accept the default, which is to add one variable.

The ***Variable*** specification dialog will be displayed. In the ***Display format*** group, select ***Number***. In the ***Long name*** field at the bottom of the dialog, enter:

=mean(v3:v25)

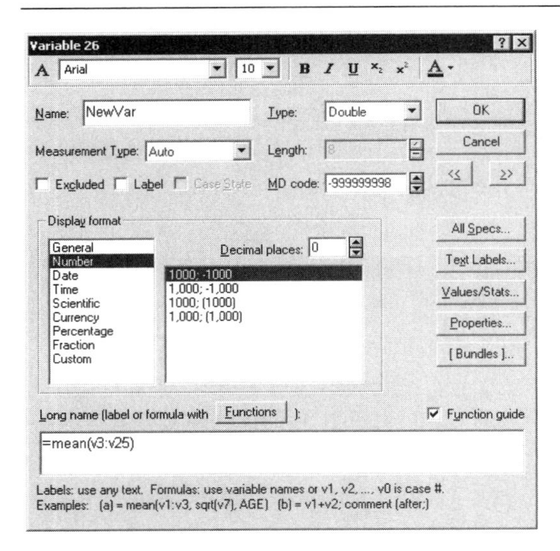

Click the ***OK*** button. A dialog will be displayed that informs you whether the formula is formally correct. Click the ***Yes*** button to proceed. The new variable is now filled with the mean of variables *3* through *25* for each case.

Since you can refer to variables by their names or their numbers; the formula we just created could also be expressed as:

=mean(MEASURE01:MEASURE23)

Example: Batch Formulas

Spreadsheet Formulas are useful for defining a formula for one variable at a time. However, there are many situations in which you need to evaluate several formulas for different variables at the same time. This can be done with the Batch Formulas facilities in *STATISTICA*.

Open the C*haracteristics.sta* data file. This data file contains information about patients in a study. For this example, we will 1) calculate patient Body Mass Index (BMI) and 2) convert height to centimeters (cm), and add these two variables to the data set.

From the ***Data*** menu, select ***Batch Transformation Formulas*** to display the ***Batch Transformation Formulas*** dialog.

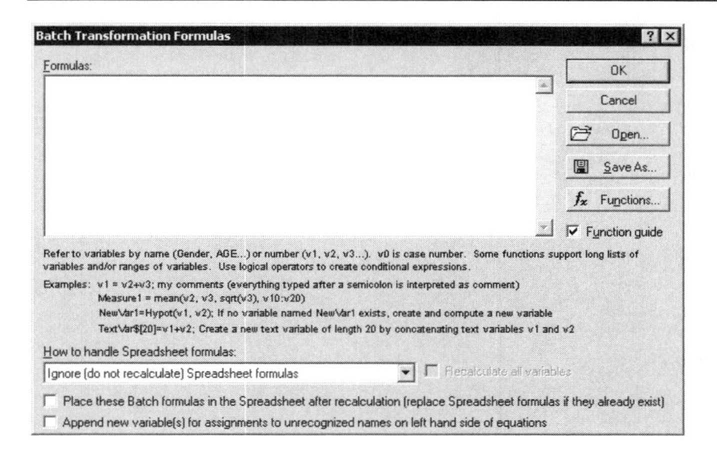

The only differences in syntax between the Batch transformation formulas and the Spreadsheet Formulas is the support for multiple formulas in the Batch option, and the fact that because the batch formulas are not attached to any specific variable (in fact they can be freely copied from data file to data file), they cannot start with an equal sign, but must have a target variable (e.g., *v1=...* or *Measure03=...*) so that *STATISTICA* knows to which variable each formula should apply. There is also an option to "distribute" all batch formulas into the respective variables in the spreadsheet and save them with the data file, effectively replacing the Spreadsheet Formulas (if there are any).

Following are the calculations used to calculate *BMI* and to convert *Height (in)* to centimeters, and the formulas to enter in the ***Batch Transformation*** dialog:

Calculation	Batch Transformation Dialog Entry
$BMI = \frac{weight(lb)}{height(in)^2} * 703$	BMI = ('weight (lb)' / 'Height (in)' **2)*703
$height(cm) = height(in) * 2.54$	'Height (cm)' = 'height (in)' *2.54

In the ***Formulas*** field, enter the list of transformation formulas to be applied to the active data spreadsheet. Separate each transformation formula by a return (press ENTER on your keyboard).

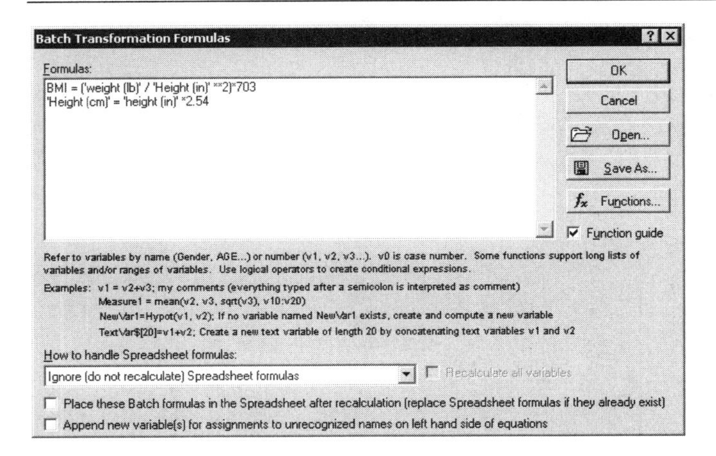

Click the ***OK*** button in the ***Batch Transformation Formulas*** dialog.

The ***Add New Variables?*** dialog will be displayed; click the ***Yes*** button to add the two new variables to the *Characteristics.sta* data file.

A dialog will be displayed to inform you whether the expressions you entered in the ***Batch Transformation*** dialog are correct. If they are OK, click ***Yes*** to proceed.

STATISTICA calculates the formulas and adds the two variables, *BMI* and *Height (cm)*, to the spreadsheet.

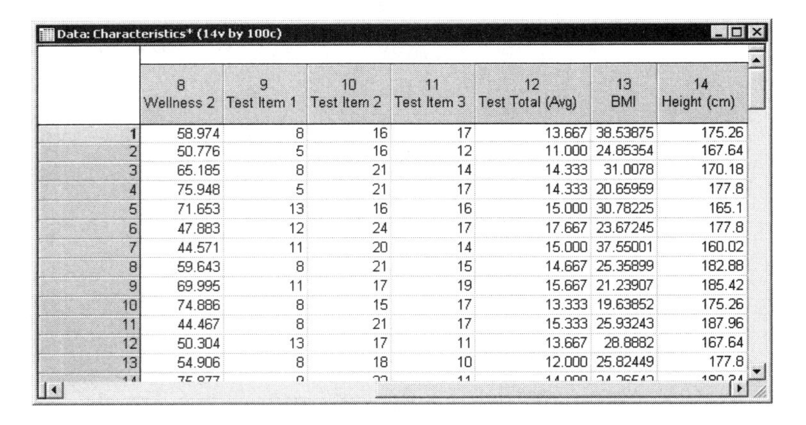

Data: Characteristics* (14v by 100c)

	8 Wellness 2	9 Test Item 1	10 Test Item 2	11 Test Item 3	12 Test Total (Avg)	13 BMI	14 Height (cm)
1	58.974	8	16	17	13.667	38.53875	175.26
2	50.776	5	16	12	11.000	24.85354	167.64
3	65.185	8	21	14	14.333	31.0078	170.18
4	75.948	5	21	17	14.333	20.65959	177.8
5	71.653	13	16	16	15.000	30.78225	165.1
6	47.883	12	24	17	17.667	23.67245	177.8
7	44.571	11	20	14	15.000	37.55001	160.02
8	59.643	8	21	15	14.667	25.35899	182.88
9	69.995	11	17	19	15.667	21.23907	185.42
10	74.886	8	15	17	13.333	19.63852	175.26
11	44.467	8	21	17	15.333	25.93243	187.96
12	50.304	13	17	11	13.667	28.8882	167.64
13	54.906	8	18	10	12.000	25.82449	177.8

The options in the ***Batch Transformation Formulas*** dialog are particularly well suited (optimized) for transforming large data sets. The formulas will be evaluated one by one, in sequence, so that any results of one transformation in the list can serve as the input for the next. Thus, it is possible to create a new variable with one formula and then use that variable in subsequent formulas.

Click the [?] button in the upper-right corner of the ***Batch Transformation Formulas*** dialog to display the *STATISTICA Electronic Manual* topic related to these options and links to various other topics containing examples of formulas and syntax rules.

EXAMPLE 5: SUMMARY RESULTS PANELS (QUALITY, PROCESS, GAGE-SIXPACKS)

Several analyses in *STATISTICA* support summary graphs and reports arranged into a single (graphics) document. In Six Sigma and manufacturing applications, these types of displays are sometimes referred to as "Quality Sixpacks," because they summarize the quality of a single variable with six (or fewer) individual graphs and tables.

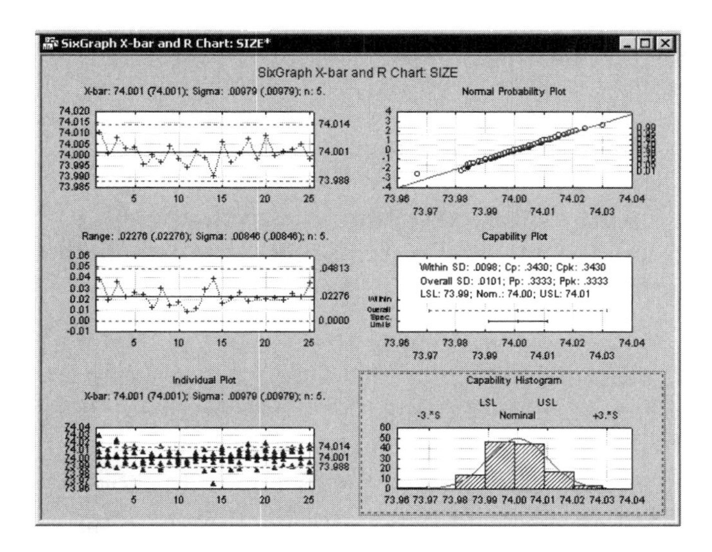

STATISTICA incorporates many such displays to summarize basic descriptive statistics, correlations, the results of Gage or process capability studies, or other types of data analyses, as shown in the following illustration.

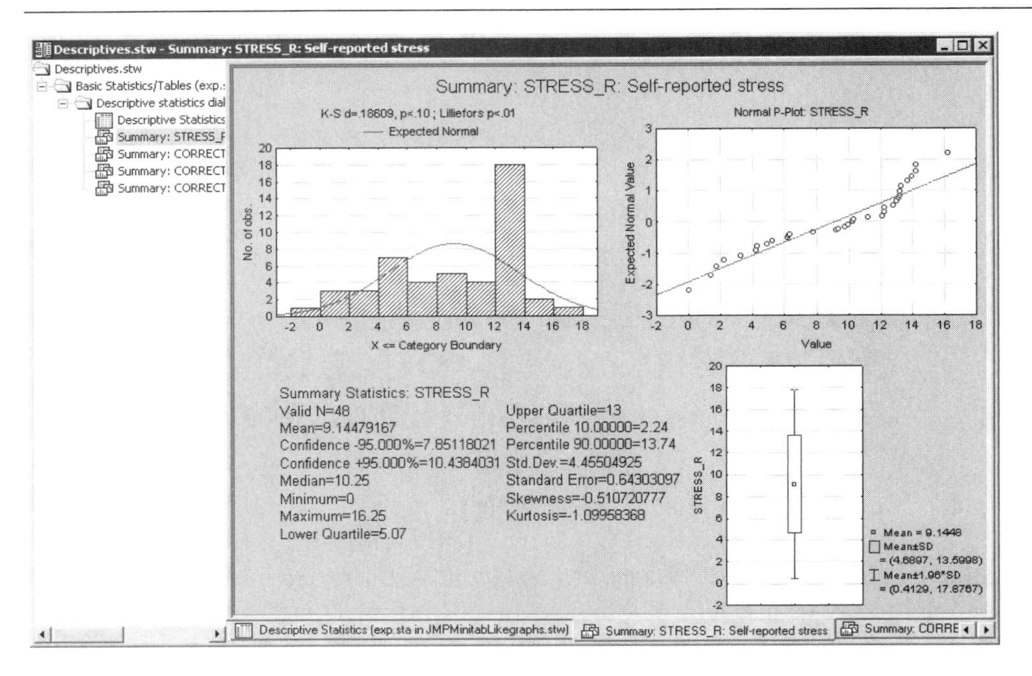

Process Capability Analysis Consistent with DIN 55319 and ISO 21747

In recent years, European (and other international) manufacturers have developed standards for the computation of process capability indices that will explicitly account for systematic and random process variation over time, as well as non-normal distributions. These indices have, for example, been adopted throughout the auto manufacturing industry and their suppliers, and *STATISTICA* fully supports these standards.

Process capability indices measure the number of times that the observed (normal) distribution of values can fit inside the specification limits for the respective part under consideration. Thus, these indices summarize the quality of a process to produce products or parts that are consistent with design specifications. In short, DIN (*Deutsche Industrie Norm*) 55319 and *ISO 21747* describe the rules to apply when choosing among various distribution models and how to account for time-dependent variation in the process.

For example, even if a distribution of data points within each sample is *Normal*, if there is systematic or random variation that occurs over time as successive samples

are taken, the resultant distribution of values will *not* be *Normal*. Therefore, in many cases the normal-distribution based process capability computations will not be applicable. Also, it is usually of interest to identify any time-dependent variability or trends, because they can indicate machine wear or other process problems.

The following example will illustrate step-by-step how to compute process capability indices consistent with these international standards, and how to create an efficient single-document summary report.

Select data. This example is based on a data set reported in Montgomery (1985, page 177, 1991, page 234). We'll use the data file *Pistons.sta* that is located in your examples directory. Specifically, we are interested in monitoring the size (diameter) of piston rings for automotive engines. Therefore, constant samples of five observations each have been taken from the ongoing manufacturing process. As is the case in many ongoing manufacturing processes, samples are taken over time, so any variability in the process quality over time will affect the overall variability.

From the ***File*** menu, select ***Open Examples*** (to open the file folder with the example data files); then open the *Datasets* folder, and double-click on *Pistons.sta* or select it and click the ***Open*** button.

Specify analysis. From the ***Statistics - Industrial Statistics & Six Sigma*** submenu, select ***Process Analysis***. In the ***Process Analysis Procedures*** dialog, select ***Process Capability ISO/DIN (Time dependent distribution model)***.

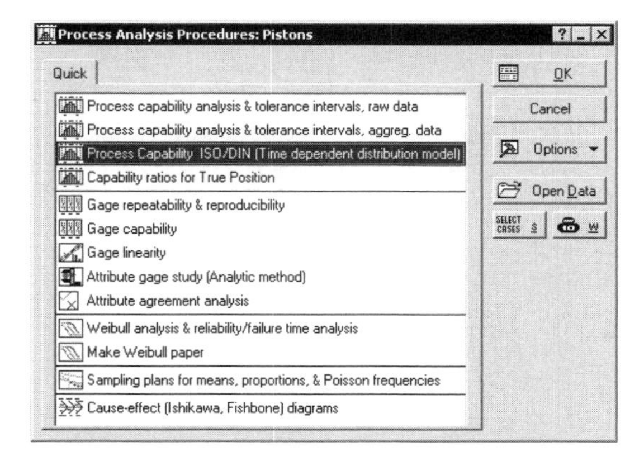

Click the ***OK*** button in the ***Process Analysis Procedures*** dialog. On the ***Quick*** tab of the ***ISO 21747 - Process Capability Setup*** dialog, click the ***Variables*** button. In the

Select Variables (and optional grouping variable) dialog, select variable *Size* in the ***Variables for the analyses*** list, and *Sample* in the ***by ... (Time/Grouping var.)*** list, and click ***OK***.

In the ***ISO 21747 - Process Capability Setup*** dialog, click the ***Process specs*** button to display the ***Enter/edit specification limits*** dialog, where you can enter the process specification limits. Specification or design limits define the maximum and (or) minimum allowable values for the respective part; in this case, specify the lower and upper spec limits (LSL, USL) as 74 +/- 0.05 (LSL=73.95, USL=74.05). Enter *74* in the ***Nominal*** field, and enter *0.05* in the ***Delta*** field.

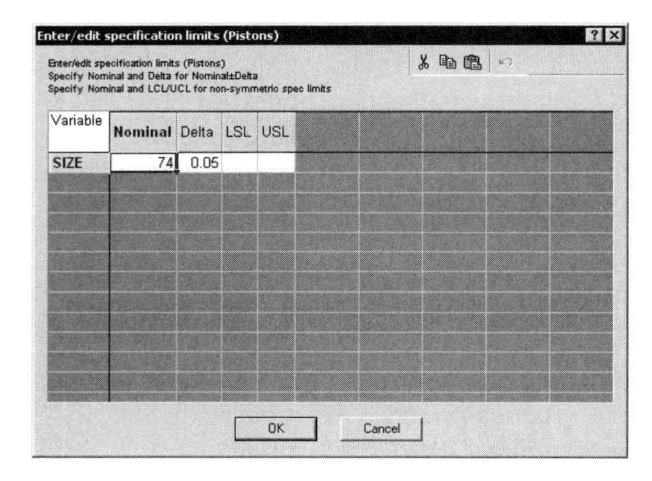

Click ***OK*** to finalize this choice.

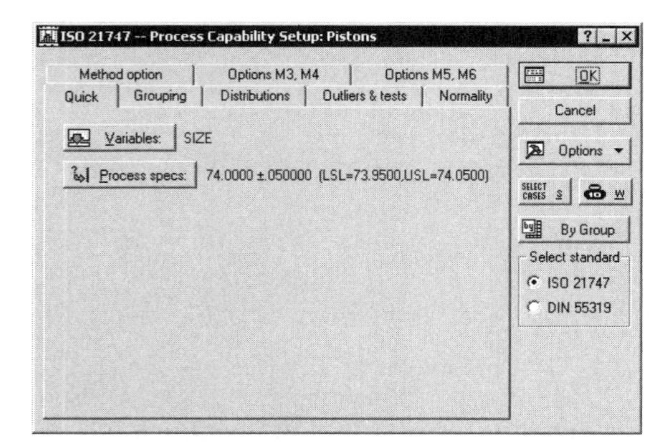

In this dialog, there are numerous other options available to modify the rules that are applied to select the most appropriate distribution and time-dependent distribution model for the data so that the appropriate process capability indices can be computed. You can click the **?** button in the upper-right corner of the dialog or press F1 to display the *STATISTICA Electronic Help* topic containing specific details regarding all options in this dialog. For example, the details regarding the (small) differences in the DIN and ISO specifications are discussed there.

Now click the ***OK*** button in the ***ISO 21747 - Process Capability Setup*** dialog to perform the analyses for variable *Size*.

Reviewing results. In the ***ISO 21747 - Process Capability Results*** dialog, click the ***Summary*** button to review the analysis summary display.

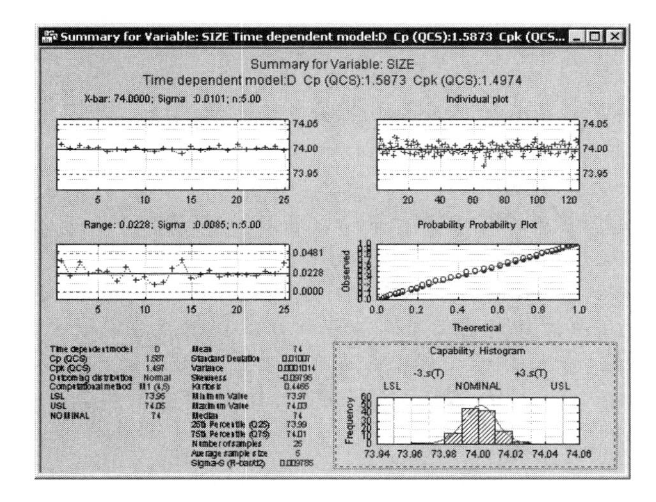

As you can see, all relevant details (as recommended in *ISO 21747*, and/or *DIN 55319*) are summarized on a single page (document), which contains all information necessary to judge the process as *capable* or *not capable* (or questionable).

Attribute Gage Analysis

For another example of this type of summary (compound) displays in *STATISTICA*, we will perform an attribute gage analysis.

In general, any measurement system used in manufacturing must be validated to ensure that the respective gages measure the quality characteristic of interest with sufficient accuracy and precision. Often, a gage of particular importance is the one

that determines whether a manufactured part is of sufficient quality to be accepted or rejected; in other words, the gage measures a simple accept/reject *attribute*.

To determine the quality of the gage, a study is periodically performed where the gage (accept/reject decision) is applied to reference parts with known deviations from the desired specifications. This process is described in the respective section of the *STATISTICA Electronic Manual*, as well as the AIAG (Automotive Industry Action Group) *Measurement System Analysis (MSA)* manual (2000).

This example illustrates the analysis described in the *MSA* manual on pages 81-86.

Select data. From the ***File*** menu, select ***Open Examples*** (to open the file folder with the example data files); then open the *Datasets* folder, and double-click on *AttributeGageStudy.sta* or select it and click the ***Open*** button. This file contains the data, already summarized to acceptance data, of the attribute gage study described in the *MSA* manual, (p. 84)

Specify analysis. From the ***Statistics - Industrial Statistics & Six Sigma*** submenu, select ***Process Analysis***. In the ***Process Analysis Procedures*** dialog, select ***Attribute gage study (Analytic method)***, and click the ***OK*** button. In the ***Attribute gage study (Analytic method)*** dialog, click the ***Variables*** button. Select *Part#* in the ***Part numbers*** list, *Reference* in the ***Reference values*** list, and *Acceptance* in the ***Acceptance/Response*** list, and then click the ***OK*** button to close this dialog and return to the ***Attribute gage study (Analytic methods)*** dialog. In the ***Tolerance limit for calculation*** group, specify *-0.01* as the ***Lower limit***, select the ***Display the other limit*** check box, and then specify *0.01* as that limit.

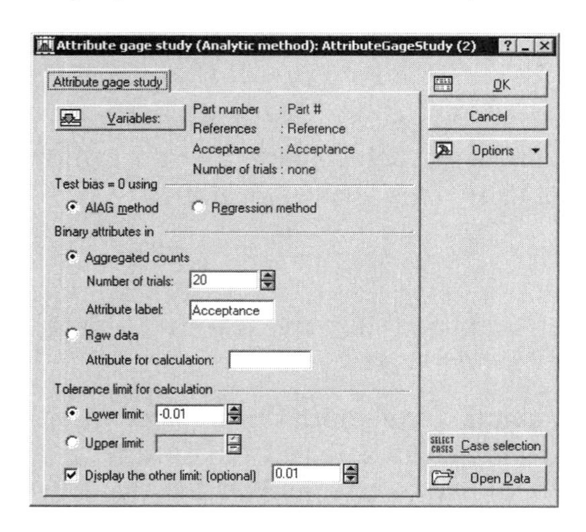

We are interested in evaluating the gage performance for a process or type of manufactured part that should be identified as unacceptable (should be rejected), when its real lower limit drops below *-0.01* (expressed here as a deviation from the spec). In the data file, the *Acceptance* probabilities summarize the number of reference parts measurements, from a total of 20 such parts and measurements each, that were declared as unacceptable (i.e., that were rejected).

Reviewing results. Now click ***OK*** in the ***Attribute gage study (Analytic methods)*** dialog. In the ***Results*** dialog, click the ***Summary*** button to review the summary results.

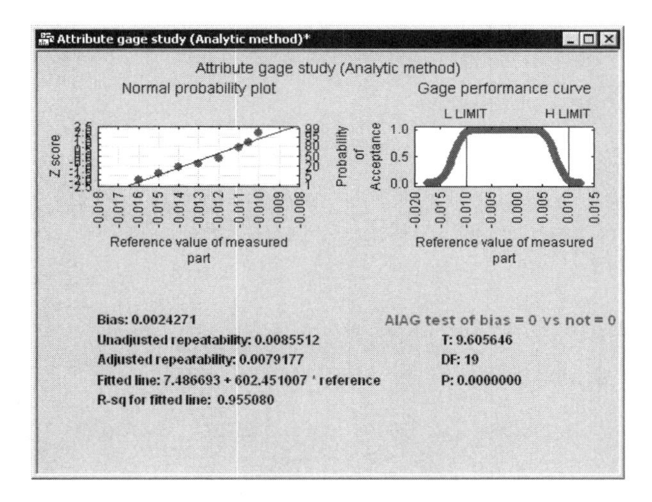

All important results to determine the bias and repeatability (of measurements) of the attribute gage are summarized on a single page. For details on the interpretation of the reported statistics and graphs, refer to the *Electronic Manual.*

EXAMPLE 6: *STATISTICA DATA MINER RECIPE*

Overview

A general trend in data mining is the increasing emphasis on solutions based on simple analytic processes, rather than the creation of ever-more sophisticated general analytic tools. The *STATISTICA Data Miner Recipe* (*SDMR*) approach provides an intuitive graphical interface to enable those with limited data mining experience to execute a "recipe-like" step-by-step analytic process. With these intuitive dialogs, you can perform various data mining tasks such as regression, classification, and clustering. Other recipes can be built quickly as custom solutions. Completed recipes can be saved and deployed as project files to score new data.

SDMR spans the entire data mining process – from querying external databases to the final deployment of solutions – and, in general, consists of the following steps.

1. Identifies the data from which to learn
    - Connects to ODBC or OLEDB compliant databases
    - Connects to *STATISTICA* data files
2. Cleans data and removes the redundant predictors
    - Flexible and efficient methods for sampling the data (simple, stratified, systematic, etc.)
    - More flexible ways to identify and recode the missing data
    - Identification of outliers
    - Transform the data prior to performing the subsequent steps
    - Identify and eliminate redundant predictors
3. Identifies important predictors from a large pool of predictors that are strongly related to the dependent (outcome or target) variable of interest
    - Feature selection for very large data sets (e.g., thousands of variables)
    - Detection of important interactions among the predictors by using tree-based methods
4. Generates a pool of eligible models
    - Leverage the comprehensive selection of cutting edge techniques for predictive data mining available in *SDMR*
    - Offload computationally expensive tasks to *WebSTATISTICA*, freeing your local computer for other tasks
5. Performs automatic competitive evaluation of models to identify the optimum model with respect to performance, and complexity
6. Deploys the model to score new data using the inbuilt efficient deployment engine

This example illustrates how quickly and efficiently data mining projects can be completed using the *STATISTICA Data Miner Recipe*, even if the best solution to the (prediction) problem emerges only after (automatically) comparing the efficacy of various advanced data mining algorithms.

Data file. In this example, we will explore the use of the *STATISTICA Data Miner Recipe* for Credit Scoring applications. The example is based on the data file *CreditScoring.sta*, which contains observations on 18 variables for 1,000 past applicants for credit. Each applicant was rated as "good credit" (700 cases) or "bad credit" (300 cases). We want to develop a credit scoring model that can be used to determine if a new applicant is a good credit risk or a bad credit risk, based on the values of one or more of the predictor variables. An additional "Train/Test" indicator variable is also included in the data file for validation purposes.

Using *STATISTICA Data Miner Recipes* (*SDMR*)

To use *SDMR* for this application, follow these steps:

Open ***STATISTICA***, and from the ***Data Mining*** menu, select ***Data Miner - Recipes*** to display the ***Data miner recipe*** dialog. Click the ***New*** button to create a new project.

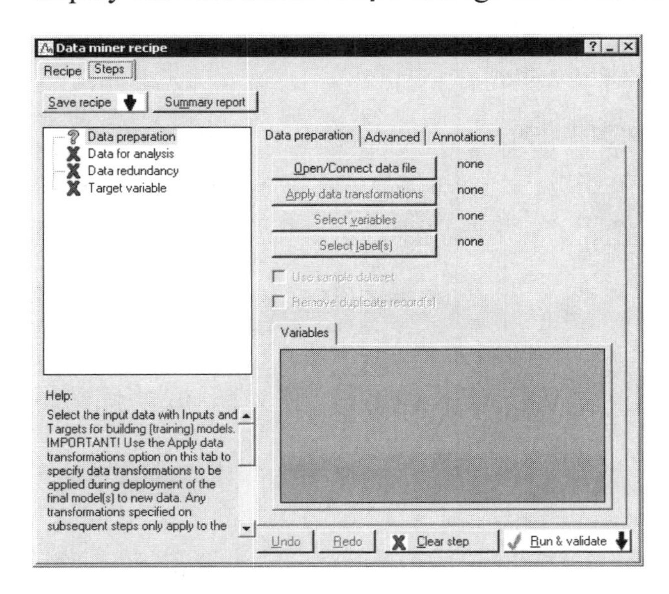

The step-node panel is located below the ***Steps*** tab. It contains four major nodes: ***Data preparation***, ***Data for analysis***, ***Data redundancy***, and ***Target variable***.

Nodes (steps). Each node (or step) can exist in one of three states at most (depending on whether its completion is optional). Each state is represented by an icon: a red ✗ indicates a wait state, meaning a step cannot be started because it is dependent on a previous step that has not been completed; a yellow ? indicates a ready state, meaning you are ready to start the step because previous steps have been completed; a green ✓ indicates a completed step. Note that you must click the ***Run & validate*** button to change the yellow ? (ready state) to the green ✓ (completed state). The change will be made only if the step has been successfully completed.

Data Preparation

Connecting data. On the ***Data preparation*** tab, click the ***Open/Connect data file*** button and double-click on the *CreditScoring.sta* data file (located in the *Datasets* folder installed with *STATISTICA*). Click the ***Select variables*** button, and from the ***Select variables*** dialog, select:

- Variable *1* (*Credit Rating*) as the ***Target, categorical*** variable,
- Variables *3*, *6*, and *14* as ***Continuous predictors***
- Variables *2*, *4-5*, *7-13*, and *15-18* as ***Categorical predictors***, and
- Variable *19* (*TrainTest*) as the ***Validation sample*** variable.

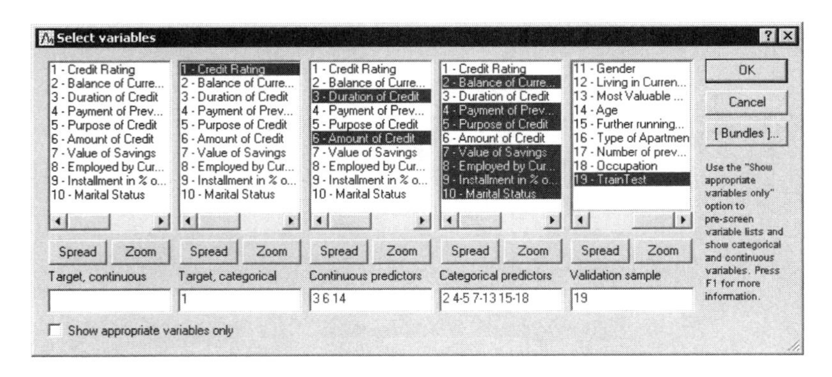

Then, click the ***OK*** button.

Select the ***Advanced*** tab in the ***Data miner recipe*** dialog, and select the ***Use sample data*** check box. Select the ***Stratified random sampling*** option button as the sampling

strategy to ensure that each class of the dependent variable *Credit Rating* is represented with approximately equal numbers of cases in train and validation sets. Then click the ***More options*** button to display the ***Stratified sampling*** dialog. Click the ***Strata variables*** button, select *Credit Rating* as the strata variable, and click ***OK*** in this dialog and in the ***Stratified sampling*** dialog.

Click the ***Run & validate*** button for the ***Data preparation*** step to ensure that this step has been successfully completed (in the step-node panel next to ***Data preparation***, the yellow ? changes to a green ✓).

Data for Analysis

After the ***Data preparation*** step is completed, the ***Data for analysis*** step will be selected automatically. On the ***Data for analysis*** tab, click the ***Select validation sample*** button, and in the ***Validation Sample Specifications*** dialog, select the ***Variable*** option button. Verify that the category (value) *Train* is entered in the ***Code for training sample*** field and *Test* is entered in the ***Code for validation sample*** field.

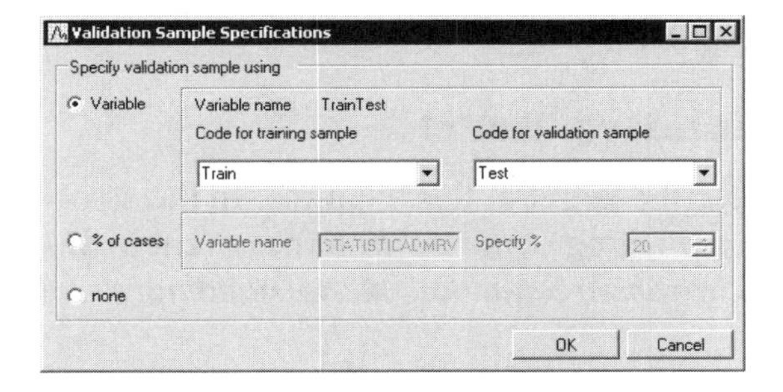

Then, click the ***OK*** button. The models will be fitted using the training sample and evaluated using the observations in the testing sample. By using observations that did not participate in the model fitting computations, the goodness-of-fit statistics computed for (predicted values derived from) the different data mining models (algorithms) can be used to evaluate the predictive validity of each model and, hence, can be used to compare models, and to choose one or more over others.

Descriptive statistics. This step will also compute descriptive statistics for all variables selected in the analysis. Descriptive stats provide useful information about ranges and distributions of the data used for the project.

Click the ***Run and validate*** button to ensure that this step is successfully complete.

Data Redundancy

Now, the ***Data redundancy*** step will be selected. The purpose of the ***Data redundancy*** step is to eliminate highly redundant predictors. For example, if the data set contained two measures for weight, one in kilogram the other in pounds, then those two measures would be redundant.

On the ***Data redundancy*** tab, select the ***Correlation coefficient*** option button, and specify the ***Criterion value*** as *0.8*. Click the ***Apply redundancy criterion*** button to eliminate the redundant predictors that are highly correlated ($r \geq 0.8$). Since there is no redundancy in the data set we are using in this example, a message dialog will be displayed stating this.

Click the ***OK*** button. Now, click the ***Run & validate*** button; the data cleaning and preprocessing for model building is now complete.

Target Variable: Building Predictive Model

Next, we need to build predictive models for the target in this example. In the step-node panel, the ***Target variable*** node has a branching structure with the parent node connecting to four child nodes including ***Dimension reduction***, ***Model building***, ***Evaluation***, and ***Deployment***.

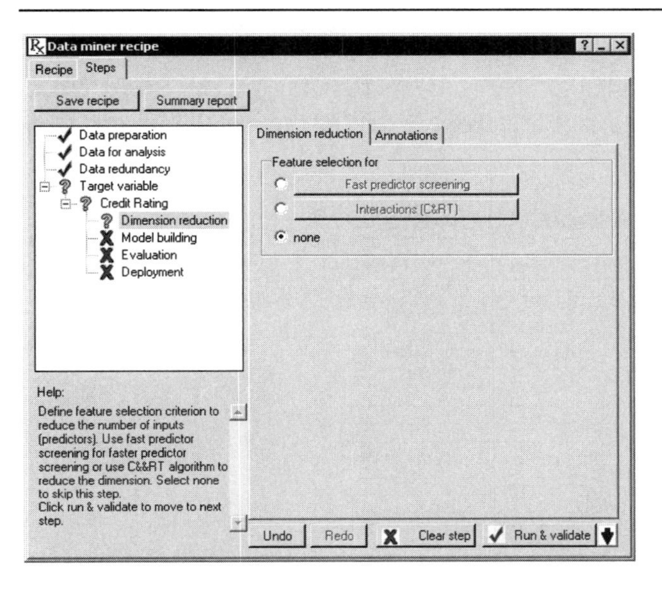

Dimension reduction. The ***Dimension reduction*** node is selected automatically. In this step, the goal is to reduce the dimensionality of the prediction problem, i.e., to select a subset of inputs that is most likely related to the target variable (in this example *Credit rating*) and, thus, is most likely to yield accurate and useful predictive models. This type of analytic strategy is also sometimes called *feature selection*.

Two strategies are available. If the ***Fast predictor screening*** option button is selected, the program will screen through thousands of inputs and find the ones that are strongly related to the dependent variable of interest. If the ***Interactions (C&RT)*** option button is selected, tree methods are used to detect important interactions among the predictors.

For this example, select the ***Interactions (C&RT)*** option button as the feature selection strategy, and then click the ***Interactions (C&RT)*** button to display the ***C&RT*** dialog. Enter *12* in the ***Number of predictors to extract*** field and in the ***Prior class probabilities*** field.

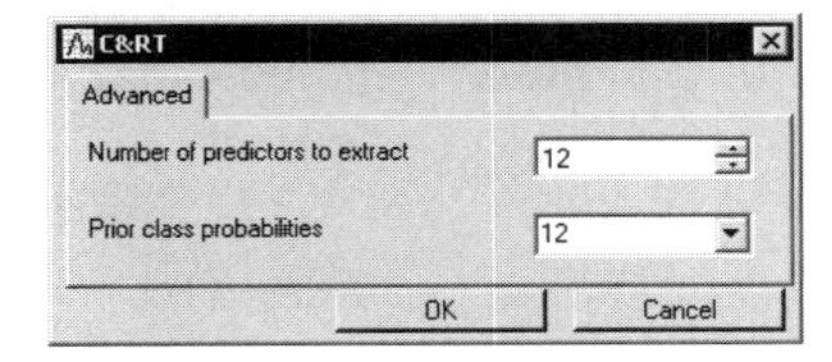

Click the ***OK*** button in this dialog, and then click the ***Run & validate*** button to complete this step. To review a summary of the analysis thus far, on the ***Steps*** tab, click the ***Summary report*** button, and from the drop-down list, select ***Summary report*** to display the *Results* workbook.

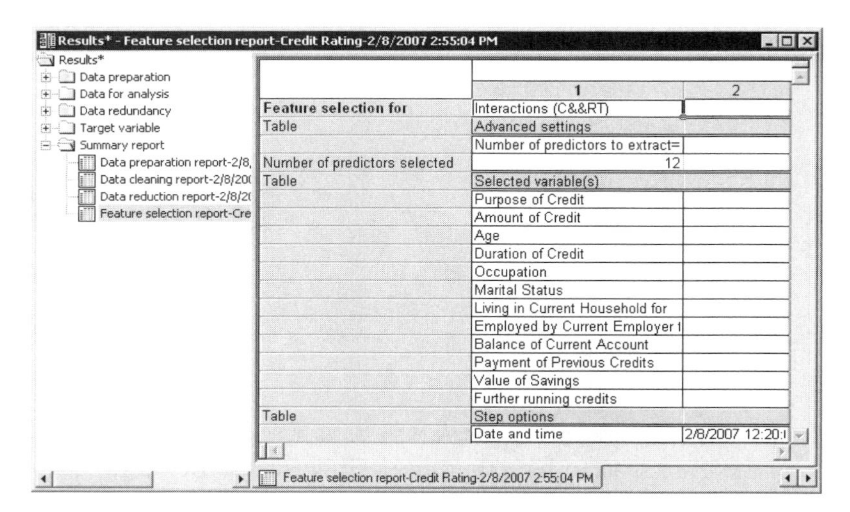

These predictors will be further examined using various cutting-edge data mining and machine learning algorithms available in *SDMR*.

Building models. The ***Data miner recipe*** dialog was minimized so that you could see the *Results* workbook. Click the ***Data miner recipe*** button located on the Analysis Bar to display the dialog again. Now, the ***Model building*** node is selected. In this step, you can build a variety of models for the selected inputs. On the ***Model building*** tab, ***C&RT***, ***Random forest***, ***Boosting tree***, ***Neural network***, and ***SVM*** are selected by default as the models or algorithms that will automatically be "tried" against the data.

The computations for building predictive models can be performed either locally (on your computer) or on the *WebSTATISTICA* Server. However, the latter option is available only if you have a valid *WebSTATISTICA* Server account and you are connected to the server installation at your site. For this example, click the ***Build model*** button to perform the computations locally on your computer. This will take a few moments; when finished, click the ***Run and Validate*** button to complete this step.

Evaluating and selecting models. Now, the ***Evaluation*** node is selected. On the ***Evaluation*** tab, click the ***Evaluate models*** button to perform the competitive evaluation of models for identifying the best performing model in terms of performance in the validation sample. Notice that the *Boosting Trees* model has the minimum error rate of 30.39%. In other words, 69.61% of the cases in the validation sample are correctly predicted by this model. Note that your results may vary slightly because these advanced data mining methods randomly split the data into subsets during training to produce reliable estimates of the error rates.

On the ***Steps*** tab, click the ***Summary report*** button, and from the drop-down list, select ***Summary report*** to display the ***Results*** workbook.

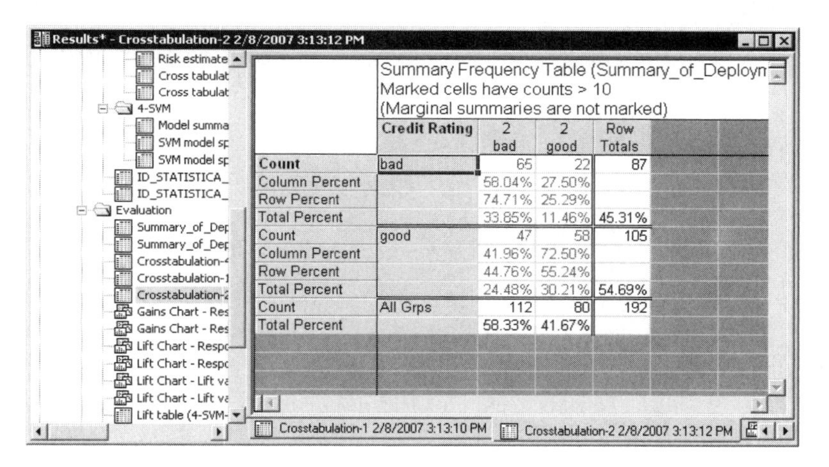
Results* - Crosstabulation-2 2/8/2007 3:13:12 PM

Summary Frequency Table (Summary_of_Deploym
Marked cells have counts > 10
(Marginal summaries are not marked)

	Credit Rating	2 bad	2 good	Row Totals
Count	bad	65	22	87
Column Percent		58.04%	27.50%	
Row Percent		74.71%	25.29%	
Total Percent		33.85%	11.46%	45.31%
Count	good	47	58	105
Column Percent		41.96%	72.50%	
Row Percent		44.76%	55.24%	
Total Percent		24.48%	30.21%	54.69%
Count	All Grps	112	80	192
Total Percent		58.33%	41.67%	

Crosstabulation-1 2/8/2007 3:13:10 PM
Crosstabulation-2 2/8/2007 3:13:12 PM

This spreadsheet shows the classification performance of the best model on the validation data set. The columns represent the predicted class frequencies, as predicted by the *Boosting Trees* model, and the rows represent the actual or observed classes in the validation sample. In this matrix, you can see that this model predicted 66 out of 87 "bad credit risks" correctly, but misclassified 22 of them. This information is usually much more informative than the overall misclassification rate, which simply tells us that the overall accuracy is 69.36%.

Display the ***Data miner recipe*** dialog again, and click the ***Run and Validate*** button to complete this step.

Deployment

The final ***Deployment*** step involves using the best model and applying it to new data in order to predict the "good or bad" customers. In this case, deploy the *Boosting*

Trees model that gave us the best predictive accuracy on the test sample when compared to the other models. This step also provides the option for writing back the scoring information (classification probabilities computed by the best model, predicted classification, etc.) to the original input data file or database. This is extremely useful for deploying models on very large data sets to "score" databases.

On the ***Deployment*** tab, click the ***Data file for Deployment*** button and double-click on the *CreditScoring.sta* data file (located in the *Datasets* folder installed with *STATISTICA*). For demonstration purposes, we are using the same data file for deployment of the best model.

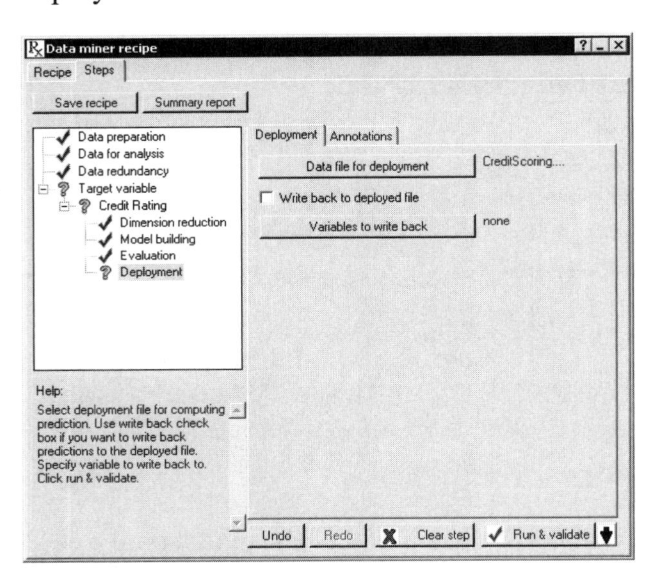

Click the ***Run and Validate*** button to score this data file using the best model. The scored file with classifications and prediction probabilities (titled *Summary of Deployment*) is located in the *Deployment* folder in the project workbook as shown below.

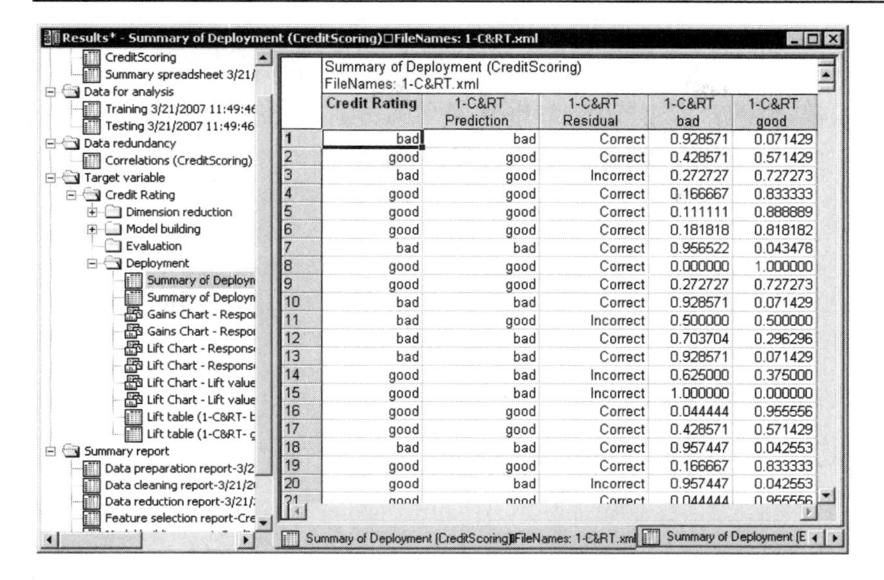

	Credit Rating	1-C&RT Prediction	1-C&RT Residual	1-C&RT bad	1-C&RT good
1	bad	bad	Correct	0.928571	0.071429
2	good	good	Correct	0.428571	0.571429
3	bad	good	Incorrect	0.272727	0.727273
4	good	good	Correct	0.166667	0.833333
5	good	good	Correct	0.111111	0.888889
6	good	good	Correct	0.181818	0.818182
7	bad	bad	Correct	0.956522	0.043478
8	good	good	Correct	0.000000	1.000000
9	good	good	Correct	0.272727	0.727273
10	bad	bad	Correct	0.928571	0.071429
11	bad	good	Incorrect	0.500000	0.500000
12	bad	bad	Correct	0.703704	0.296296
13	bad	bad	Correct	0.928571	0.071429
14	good	bad	Incorrect	0.625000	0.375000
15	good	bad	Incorrect	1.000000	0.000000
16	good	good	Correct	0.044444	0.955556
17	good	good	Correct	0.428571	0.571429
18	bad	bad	Correct	0.957447	0.042553
19	good	good	Correct	0.166667	0.833333
20	good	bad	Incorrect	0.957447	0.042553

Summary

The purpose of this example is to demonstrate the efficiency of the data miner workflow implemented in the *STATISTICA Data Miner Recipe.* With only a few clicks, the program will take you through the complete analytic process - from the definition of input data and analysis problem, through data cleaning and preparation and model building, all the way to final model selection and deployment.

Even though most of the computational complexities of data mining are resolved automatically in the *STATISTICA Data Miner Recipe*, which enables you to move from problem definition to a solution very quickly even if you are a novice, the program will "apply and try" a large number of advanced data mining algorithms and automatically determine which approach is most successful.

Thus, the *STATISTICA Data Miner Recipe* methodology and user interface enables you to leverage the largest collection of data mining algorithms in a single package to solve your problems.

EXAMPLE 7: *WebSTATISTICA* – DOWNLOAD/OFFLOAD ANALYSES FROM/TO SERVERS

WebSTATISTICA Server extends the capabilities of the *STATISTICA* platform, turning several standalone workstations into a powerful, enterprise-wide collaborative-intelligence system. One of the key features of *WebSTATISTICA*'s client-server architecture is that it enables you to utilize server-side resources to run multiple, possibly time-consuming or repetitive statistical analyses ("offload" tasks to the server) while at the same time freeing the local system for other tasks that require immediate attention. This can be achieved using either a Web browser ("thin" client) or desktop version of *STATISTICA* ("thick" client, *WebSTATISTICA* client).While the former allows access to *WebSTATISTICA* Server from virtually any computer in the world, the latter requires *STATISTICA* installation on your computer. *WebSTATISTICA*'s tight integration with the *STATISTICA* application provides common user experience and workflow for both client and server-side operations, a generally more feature-rich and responsive user interface, and all the additional components and tools of desktop *STATISTICA*.

Offloading an analysis (or a custom script) to *WebSTATISTICA* Server. First, make sure that *WebSTATISTICA* integration is enabled. From the *STATISTICA* ***Tools*** menu, select ***Options*** to display the ***Options*** dialog. Click on the ***Server / Web*** tab. Select the ***Enable WebSTATISTICA Server Integration*** check box. The only required parameter is *WebSTATISTICA* Server's network path (and connection settings, if they are different from the default). Ask your network administrator for these values. It is possible to ***Enable Integrated Login***, if it is supported and enabled on the server; otherwise you will need to enter your user name and password when logging in to *WebSTATISTICA* Server.

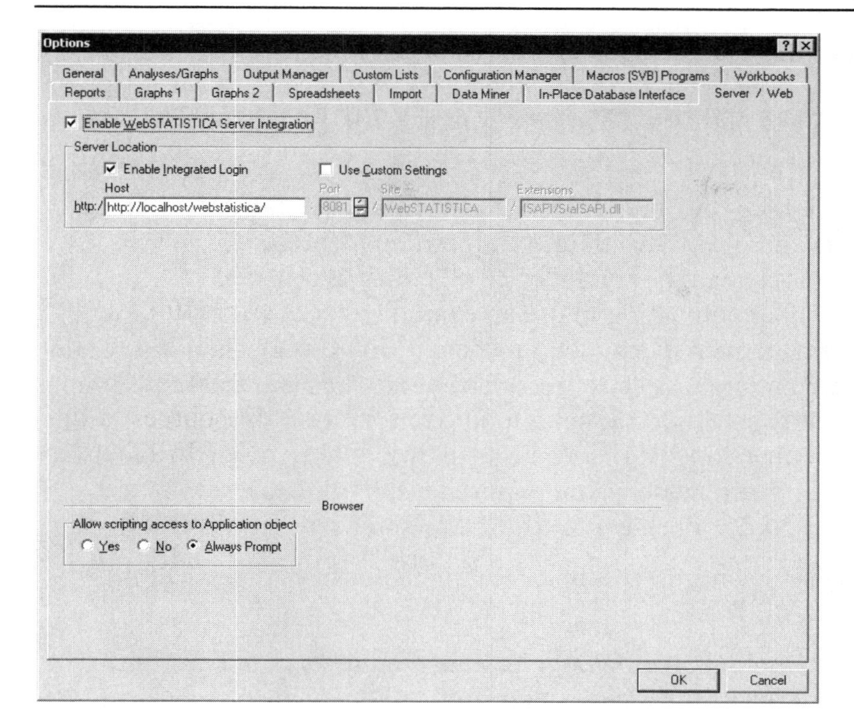

After specifying the options on this tab, click the ***OK*** button.

The ***Server*** menu has now been added to the *STATISTCA* toolbar. From the ***Server*** menu, select ***Log In***, and enter your user name and password if requested. Upon successfully establishing a connection, the menu commands will become available.

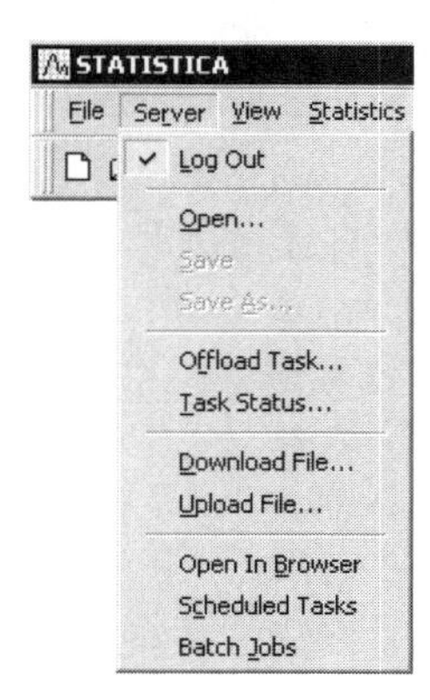

The ***Open***, ***Save***, and ***Save As*** commands on the menu are used to upload a currently open file to the server or download a file and open it locally. There are also explicit commands to ***Upload File*** to and ***Download File*** from specific folders on the server and the client.

Note: As real-world examples of time- or resource-consuming analyses are usually based on large data sets and/or involve iterative algorithms represented by *STATISTICA* components that are not included in all configurations of *STATISTICA*, we are deliberately going to use an example that does not require much time to complete. But even in a situation where a single analysis is quick and not resource-intensive, you might need to run a fairly complicated, time-consuming sequence of tasks, possibly scheduled at certain time intervals. In this case, *WebSTATISTICA* scheduling facilities could be used once you have created and uploaded a custom script that represents the required tasks (for example, by combining the macros recorded during a *STATISTICA* session).

Record a sample analysis macro, for example, complete the steps described in *Example 2: ANOVA* (page 34).

After completing the example, in the ***ANOVA Results*** dialog, click the ***Options*** button (located at the bottom of the dialog), and from the drop-down list, select ***Create Macro***. In the ***New Macro*** dialog, accept all defaults, and click ***OK***. Test the generated macro by running it (press F5) to make sure that it produces results as expected. Click on the macro code window to make sure it has the focus.

Then, from the ***Server*** menu, select ***Offload Task*** to display the ***Offload a task*** dialog.

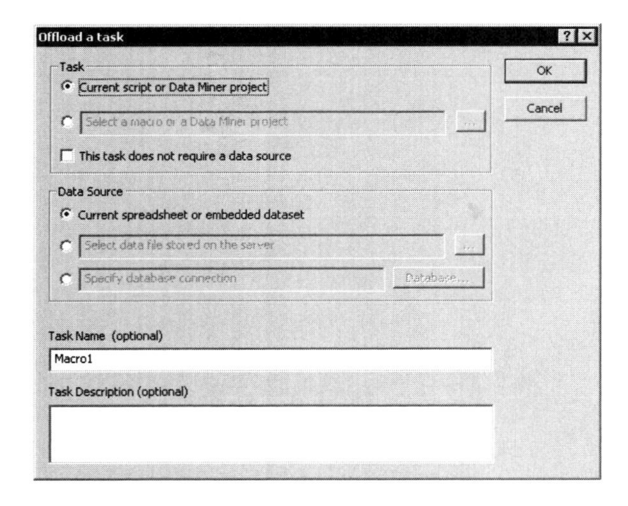

We need to select a task to offload (a script or a *Data Miner* project) and, optionally, a data set on which the task will operate (the data set could be an optional component since *Data Miner* projects may have their data sets embedded and macros might explicitly load data sets or not require them at all).

Since there is an open active data set (*Adstudy.sta*) and an open *STATISTICA* macro (our sample analysis), the default settings of the options in the ***Offload a task*** dialog specify to use them for offloading. Instead, this example will demonstrate how to reference a task and a server-side data set. This option is useful since it gives you the advantage of central server-side storage, which is especially beneficial in the case of large data sets (possibly dynamically updated) that are used by multiple users.

To reference a server-side data set, in the ***Data Source*** group box, select the ***Select data file stored on the server*** option button to display the ***WebSTATISTICA Repository*** dialog.

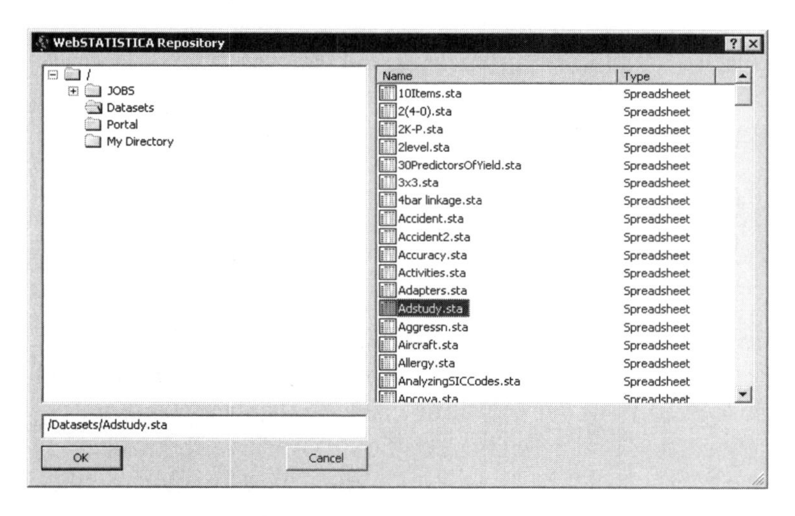

The directory structure in the tree view of the dialog represents the *WebSTATISTICA* Server Repository (possibly abridged according to your particular permissions). Click on the *Datasets* folder in the left pane, and select *Adstudy.sta* in the right pane (or you can enter the path in the edit box at the bottom of the dialog).

Click ***OK*** in the ***WebSTATISTICA Repository*** dialog and in the ***Offload a task*** dialog. *STATISTICA* will submit the task to the server, uploading files if needed. Now you can switch to other activities, while periodically monitoring the status of offloaded tasks by selecting ***Task Status*** from the ***Server*** menu to display the ***Task Status***

dialog. The following illustration shows a ***Task Status*** dialog containing several offloaded tasks.

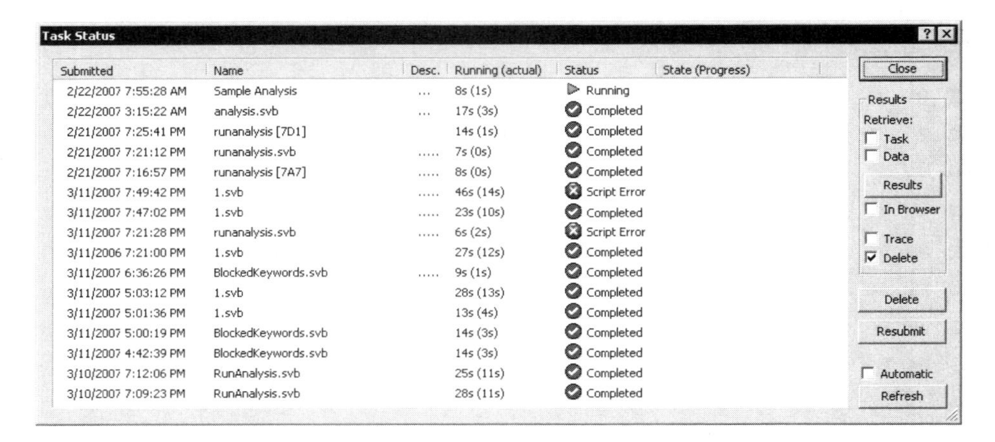

The task list status can be updated manually by clicking the ***Refresh*** button or automatically by selecting the ***Automatic*** check box in the lower-right portion of the ***Task Status*** dialog. Tasks go through ***Pending*** and ***Running*** states to either ***Completed*** or ***Script Error***.

If your task fails, double-click on the task entry to view additional information about the failure. When the error is fixed (e.g., SVB script or *Data Miner* workspace is updated), select the failed task and click the ***Resubmit*** button.

Once the task completes successfully, you can retrieve the results. Note that since the results are located on the server, they are available from any *STATISTICA* client workstation, as long you are logged in under the same credentials. The ***Results*** group box contains a ***Task*** check box and a ***Data*** check box to retrieve the task source and the data set (if applicable) back to the client. When the ***In Browser*** check box is selected, the results will be opened in the browser, switching to a thin client. This option is useful if the results are expected to be significant in size; e.g., if the analysis generates many data sets and/or graphs, you can search through them in the browser and select only the specific results you want to retrieve to your desktop. ***Trace report*** provides a diagnostic report of task execution.

Since the server does not "know" when the task results are no longer needed, it is up to you to delete them. Therefore, to save disk space on the server, a message will be displayed every time results are requested asking if the results should be deleted after retrieval (unless the ***Delete task after retrieval*** check box is cleared).

Once our task completes, we retrieve the results and close the ***Task Status*** dialog. Results are equivalent whether run locally or on the server.

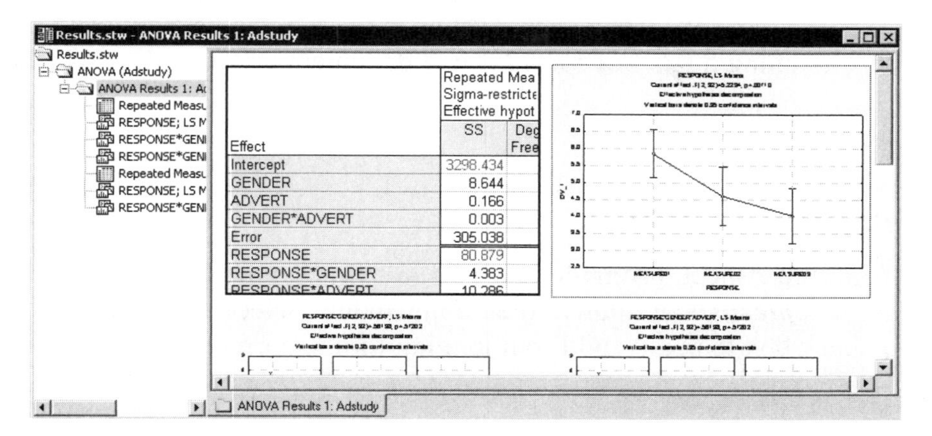

EXAMPLE 8: USING *STATISTICA* IN REGULATED ENVIRONMENTS

In a regulated environment, analyses conducted for GxP (Good Manufacturing Practices, Good Clinical Practices, Good Laboratory Practices) applications are ones that impact consumer safety such as in clinical trials, manufacturing, and quality control. When a business conducts analyses for a GxP application, regulatory bodies recommend that the company be able to prove that the results of the validated analysis system (e.g., *STATISTICA*) are under control. *STATISTICA*, through its audit trail and spreadsheet/report locking features, offers the tools you need to meet this regulatory requirement.

In order to meet traceability requirements for GxP applications, there are at least three concerns: 1) control of the input data being submitted to the analysis (i.e., knowing who made what change, at what time, for what reason; and the old values and new values), 2) control of the results tables and graphs (e.g., demonstrate that they were not altered in any way after they were created), and 3) traceability between the version of the input spreadsheet and the results outputs. *STATISTICA*

provides this information through its Spreadsheet Audit Trails and GxP Reports functionality.

See also *STATISTICA Document Management System* in the *Electronic Help* for more details about versioning/history of *STATISTICA* documents.

Control of Input Data

Enable Audit Trail Logging

Open a *STATISTICA* spreadsheet. From the ***Tools - Audit Trail*** submenu, select ***Settings*** to display the ***Spreadsheet Audit Log Settings*** dialog. Select the ***Enable audit trail logging*** check box to begin audit trail logging for the current spreadsheet.

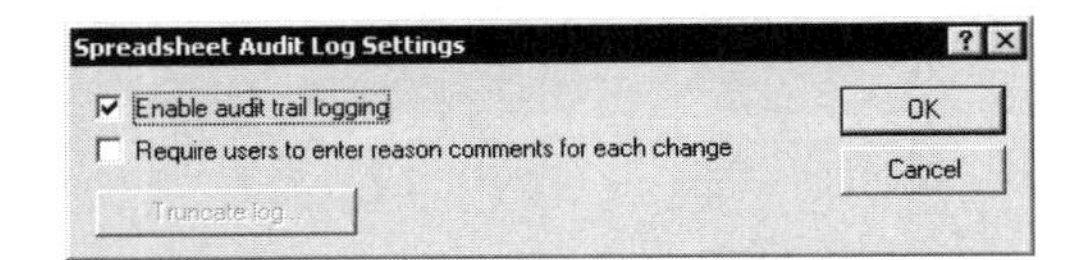

Note that when spreadsheet audit trail logging is enabled, the spreadsheet is automatically set to direct mode, i.e., changes made to the spreadsheet will be immediately written to disk. Thus, when audit trail logging is enabled, changes to the data file cannot be undone.

Select the ***Require users to enter reason comments for each change*** check box to require users to explain each change made to the spreadsheet.

The ***Truncate log*** button is available only if audit trail logging has previously been specified, and there is a current *Spreadsheet Audit Log Viewer* attached to the spreadsheet. Clicking this button will truncate the spreadsheet log and delete all existing entries. You will be prompted to confirm this action before the current entries are deleted. Once the log is truncated, the truncate action will be recorded in the newly truncated log file.

Click ***OK*** in the ***Spreadsheet Audit Log Settings*** dialog, and audit trail logging will be enabled; in fact, the ***Enter reason for change*** dialog will be displayed immediately in order to enter the reason for enabling the logging function. Enter a comment, and click ***OK***. Now, right-click in the header of the last variable in the spreadsheet, and select ***Add variables*** from the shortcut menu. In the ***Add Variables*** dialog, we will accept all defaults, so click ***OK***. The ***Enter reason for change*** dialog will be displayed; you must enter a comment and click ***OK*** before the change will be made. When audit trail logging is enabled, every change made to the spreadsheet will be documented, and when the ***Require users to enter reason comments for each change*** check box is selected, user comments also will be stored and displayed in the *Spreadsheet Audit Log Viewer*.

Next, from the ***Tools - Audit Trail*** submenu, select ***View Log*** to display the *Spreadsheet Audit Log Viewer*.

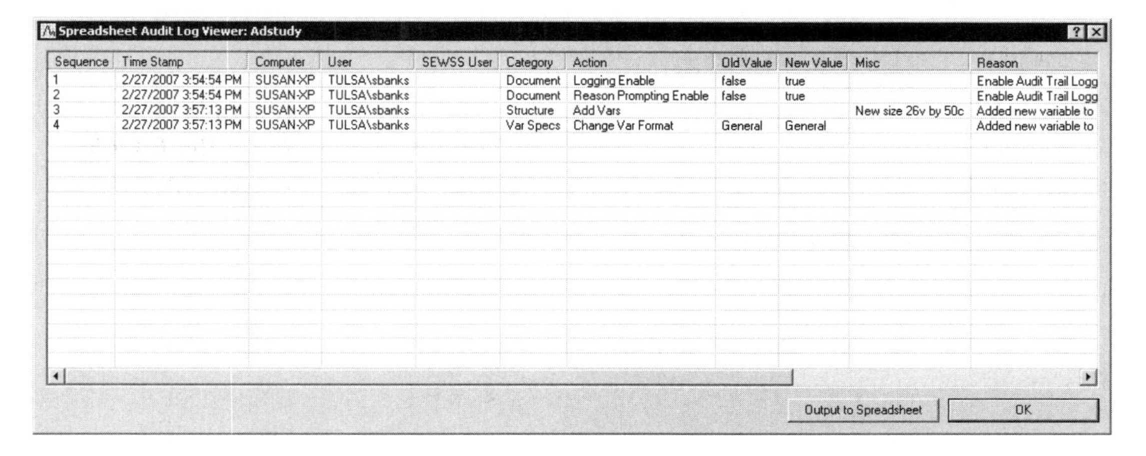

The log viewer displays a grid of information regarding the audited actions including the sequence number, time of change, the computer used to make the change, user information, the nature of the change, and the reason for the change. Column widths in the log grid can be increased and decreased using standard Windows techniques. The Spreadsheet Audit Trails are saved and embedded into each respective spreadsheet.

Password encryption vs. locking. A spreadsheet can be password encrypted so that it cannot be opened without the correct password. Only users who know the password can open the spreadsheet. Once a password encrypted spreadsheet is opened, it can be modified. Alternatively, spreadsheet locking makes portions of the spreadsheet read-only, enabling you to prevent changes to some or all aspects of the spreadsheet. The spreadsheet can be opened by anyone, but locked portions cannot be altered. Both the password encryption options and spreadsheet locking facilities can be used simultaneously.

Password Encrypt a Spreadsheet

Open a *STATISTICA* spreadsheet. From the ***File*** menu, select ***Properties*** to display the ***Document Properties*** dialog. Select the ***Password*** tab.

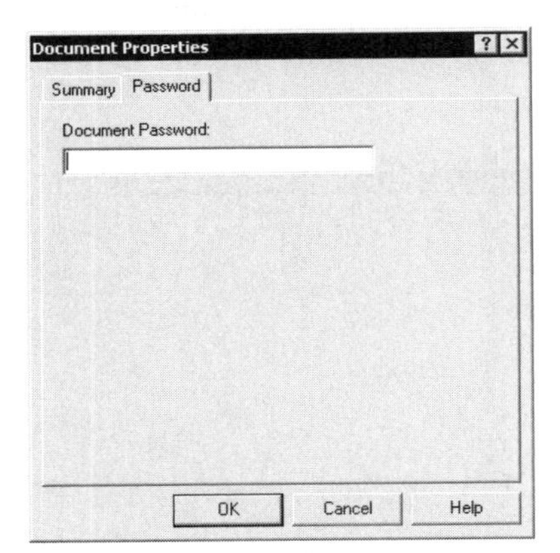

Enter a password in the ***Document Password*** field, and click the ***OK*** button. The ***Password*** dialog will be displayed, where you reenter the password to confirm it; passwords are context-sensitive.

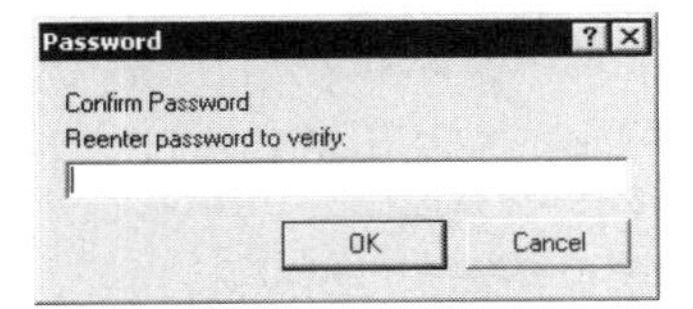

Click the ***OK*** button in the ***Password*** dialog, and close the data file. A dialog is displayed where you can choose to save the changes; click the ***Yes*** button so that the password will be encrypted. The next time anyone attempts to open this spreadsheet, the ***Password*** dialog will be displayed, and the correct password must be entered before the spreadsheet will open.

Lock a Spreadsheet

In order to meet compliance requirements, it is necessary to have control of the reliability of input data. Using the spreadsheet locking options, you can prevent changes to all spreadsheet features, from the appearance of the data (i.e., display elements, variable specifications) to the actual data and any case selection conditions or weights that are defined for the spreadsheet. Of course, sometimes

changes have to be made (e.g., when data are incorrectly entered). The *STATISTICA* Spreadsheet Audit Trail facility, when enabled, will record each change made to the spreadsheet.

With *STATISTICA* Enterprise products, only users with System Administrator permissions can modify Spreadsheet Audit Trail settings. For more information, see the *Electronic Help* for *STATISTICA Enterprise* facilities.

From the Spreadsheet ***Tools*** menu, select ***Locking*** to display the ***Lock Spreadsheet*** dialog.

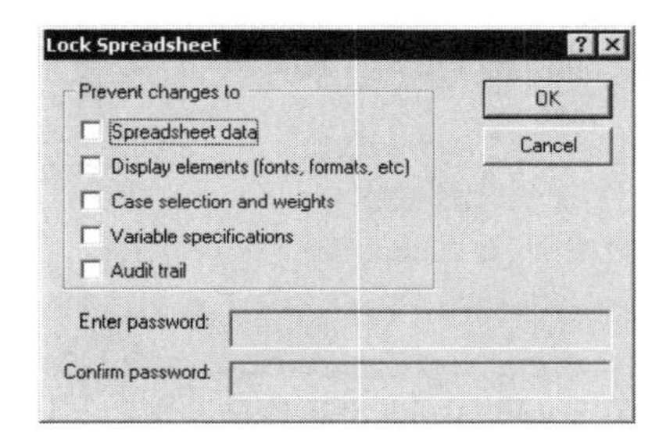

Here, you can specify which aspects of the spreadsheet that you want to lock. When users try to change a locked feature, a message will be displayed, informing them that the spreadsheet is locked.

Select the ***Spreadsheet data*** check box to prevent changes to the actual data contained in the spreadsheet. Users will be unable to change the data values and the missing data code. They will also be unable to perform any data management operations that affect the spreadsheet (e.g., change the data type or the length for text variables). If this check box is cleared, users will be able to edit the data (e.g., by updating queries and Spreadsheet Formulas or by simply typing in new values).

Select the ***Display elements (fonts, formats, etc.)*** check box to prohibit the modification of fonts and formats used in the spreadsheet. Options for changing the font size, color, type, and style (i.e., bold, underline) will be dimmed. Additionally, the options for applying spreadsheet layouts (accessible by selecting ***Layout Manager*** from the ***Format - Spreadsheet*** menu) will be unavailable.

Select the ***Case selection and weights*** check box to prevent users from changing case selection conditions and case weights for the locked spreadsheet. Users will not be able to toggle the use of selection conditions or change the currently defined selection conditions. Most options on the ***Selection*** tab of the ***Spreadsheet Case***

Selection Conditions dialog will be dimmed; however, options on the other tabs of that dialog (e.g., creating subsamples, applying formats to selection conditions) may still be available. Options on the ***Case Weights*** dialog will be unavailable.

Select the ***Variable specifications*** check box to prevent changes to the variable specifications (e.g., measurement type, missing data code, display format, long variable name). Users will still be able to view the individual variable specification dialog (accessible by double-clicking the variable header) and the Variable Specifications Editor; however, options for changing these specifications will be dimmed.

Select the ***Audit trail*** check box to prevent changes to the audit trail settings. Users will be unable to modify the audit trail settings.

Enter a password to use when locking and unlocking the spreadsheet, confirm the password (which is context sensitive), and click ***OK***. Although a password is not required, it is strongly recommended. If a password is not entered and confirmed, then any user can unlock spreadsheet features by simply clearing the selected check boxes. Note that if locks have already been defined, you must enter the correct password before locks can be changed or modified.

Now try making changes in the spreadsheet; a message will be displayed informing you that the operation cannot be completed because the spreadsheet is locked.

Controlling Results and Traceability

To meet compliance requirements, another step is to ensure that reported results are under control. *STATISTICA* provides options for creating GxP reports. In GxP mode, all results are sent to a report window, and the window is locked. All options for removing results (***Cut***, ***Extract - Original***, ***Clear***, etc.) and adding results (***Paste***, ***Insert***) are disabled. *STATISTICA* can also include a creation date in all reports as well as a time stamp for all results that are added from results dialogs. The appearance and content of the creation date and time stamp are completely configurable and can include user and computer information in addition to the time and date. Thus, in GxP mode, you can know when the results were created and by whom. You can also be certain that results have not been removed.

An additional feature of GxP mode is a traceability option. When running in GxP mode, *STATISTICA* automatically verifies whether spreadsheet audit trails are

enabled. If they are, *STATISTICA* includes the spreadsheet name and version number in the report. Sometimes version numbers are not available, for example, if audit trails are not enabled or the results are created from an In-place Database connection. When that is the case, *STATISTICA* will provide an explanation for why a version number is not available.

Create a GxP Report

From the ***Tools*** menu, select ***Options*** to display the ***Options*** dialog. Select the ***Output Manager*** tab. From the ***Report Output*** drop-down list, select either ***Send to Multiple Reports (one for each Analysis/Graph)*** or ***Single Report (common for all Analyses/Graphs)***. When one of these options is selected, the ***Report Locking (GxP Reports)*** options become available.

Select the ***Locked*** check box to ensure that documents cannot be removed from the report. Commands to ***Cut***, ***Paste***, ***Clear***, ***Rename*** and view ***Properties*** will be dimmed on the report shortcut menu. The ***Extract - Original*** command and all ***Insert*** commands will also be unavailable.

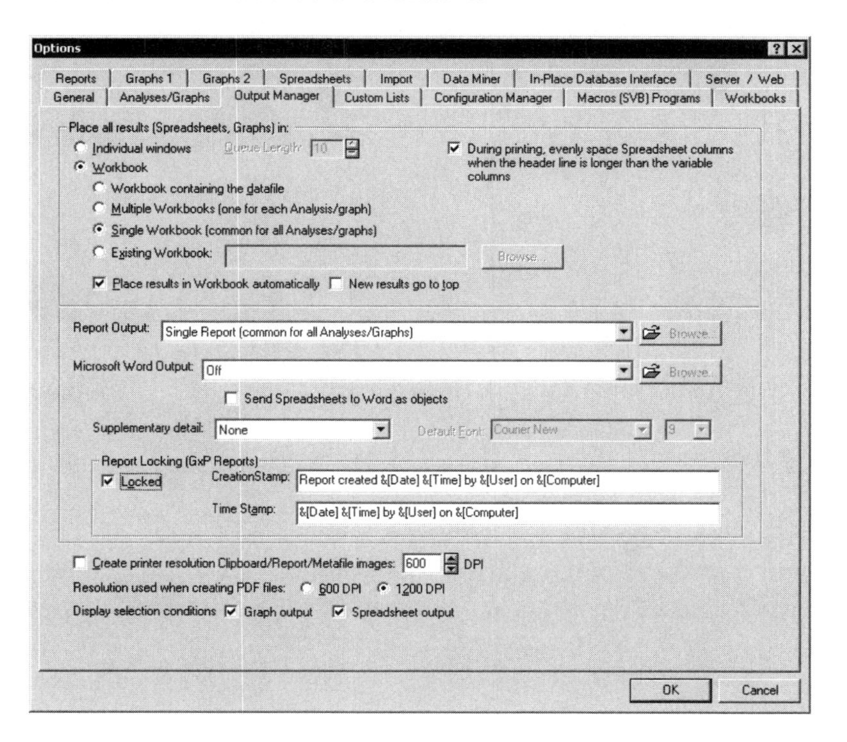

To include a creation stamp at the top of the file, you can accept the default format in the ***CreationStamp*** field, or enter your own. The following codes can be used in this field: *&[Date]*, *&[Time]*, *&[User]*, and *&[Computer]*. Any text you enter will be displayed as is.

To include a time stamp above each object as it is added to the report, you can accept the default format in the ***Time Stamp*** field, or enter your own. The following codes can be used in this field: *&[Date]*, *&[Time]*, *&[User]*, and *&[Computer]*.

Click ***OK*** in the ***Options*** dialog, and now perform any analysis, e.g., use ***Basic Statistics*** to create a quick ***Descriptive Statistics*** summary spreadsheet. When you click the ***Summary results*** button, the results will be sent to a locked report that lists the creator, date, time, etc., of the analyses.

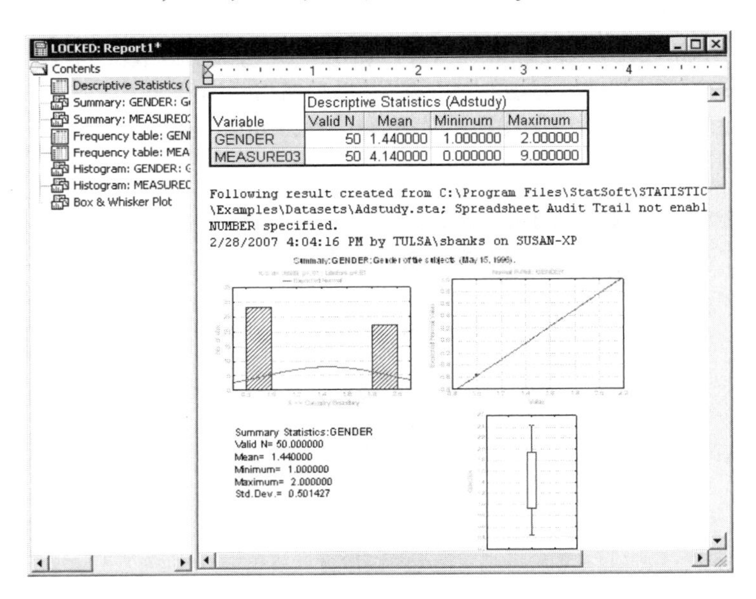

EXAMPLE 9: *STATISTICA ENTERPRISE*

STATISTICA Enterprise products extend the functionality of *STATISTICA* applications by offering collaborative work, central administration, system level customization, and other features necessary when using *STATISTICA* applications as part of the enterprise-level computer systems.

STATISTICA Enterprise Manager is a component of the *STATISTICA Enterprise* system that enables users to configure various aspects of the *Enterprise* system including user administration, system view organization, database connection maintenance, data configurations, and analysis configurations.

For this example, we will:

1. Create a new user
2. Create a new group
   a. Assign permissions to the group
   b. Add the user (see No. 1) to the group
3. Create a system view node
4. Create a new database connection
5. Create a data configuration
6. Create an analysis configuration
7. Run the analysis configuration

System View vs. Object View

Before starting this example, one thing should be noted. From the *STATISTICA Enterprise Manager* ***View*** menu, you can select either ***System View*** or ***Object View***. In ***System View***, objects, e.g., Data Configurations and Analysis Configurations, are shown as child nodes. In ***Object View***, objects are shown as child nodes within their respective categories. For this example, ***System View*** should be selected.

1. Create a New User

Launch the *Enterprise Manager*, and log in as a user who is part of the default Administrator group. In the tree view, click the plus sign ⊞ next to the *User Administration* node to expand it, and then select the *Users* folder. In the property page (the right pane), click the ***New User*** button to display the options to create a new user. In the ***Name*** field, enter *Test User 1*, and define a password and confirm the password.

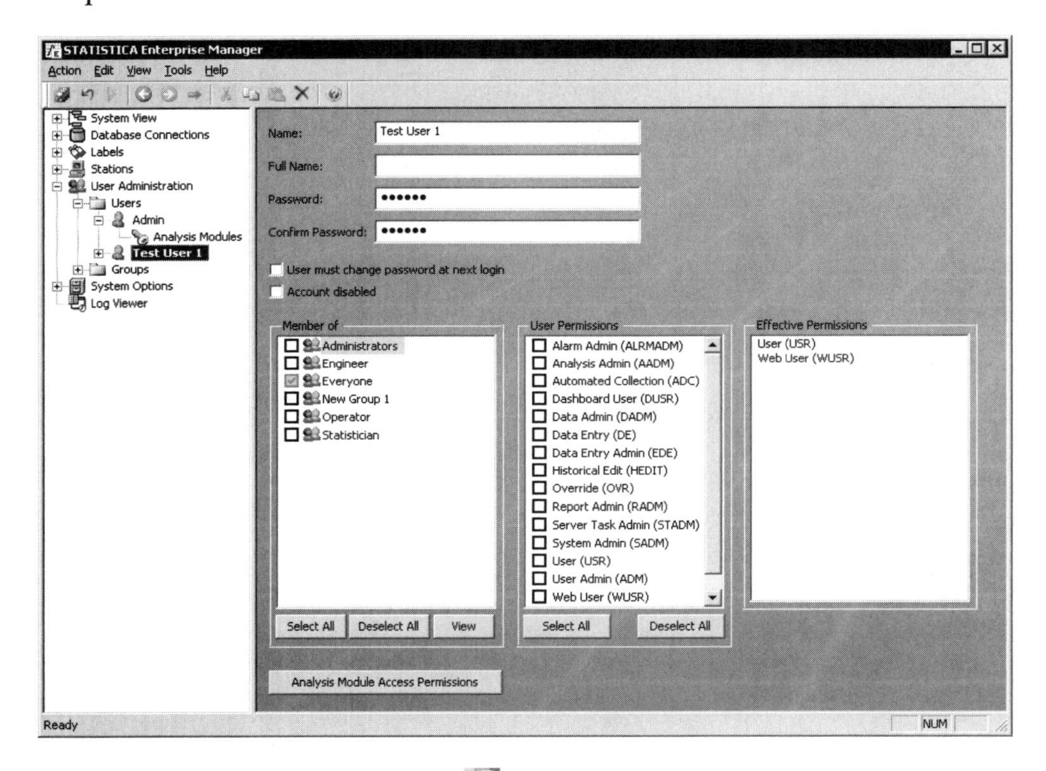

Then, click the ***Commit Changes*** toolbar button to commit the change.

We will now create a group, give the group permissions, and assign the new user to that group to allow the user to have permission to log on to the *Enterprise Manager*. With this method, any permission changes will only need to be applied to the group instead of the individual users making maintenance of users in *STATISTICA Enterprise* easier.

2. Create a New Group

In the *User Administration* node, select the *Groups* folder, and in the property page, click the ***New Group*** button to display the options to create a new group. In the ***Name*** field, enter *Test Group 1*. In the ***Group Members*** frame, select the check box adjacent to *Test User1*. This will add the previously created user to the group. In the ***Group Permissions*** frame, select the check boxes adjacent to *Analysis Admin* and *Web User*. In the tree view, click the plus sign ⊞ adjacent to the *Test Group 1* node to expand it, and select *Analysis modules*. Click the *Select All* button to select all of the modules in the ***Available analysis modules*** list.

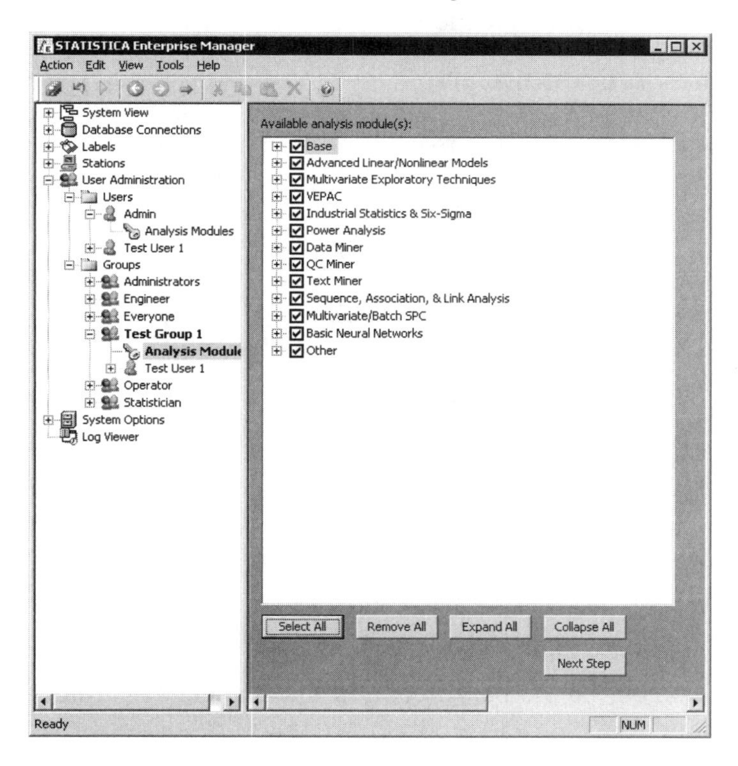

This will give users of this group permission to log on to both Web and desktop *STATISTICA* and run all of the available analyses and reports.

Click the ***Commit Changes*** toolbar button to commit the change.

We have now created the necessary user and group security to run analyses and reports. When creating the data, analysis, and report configurations in the next steps, we will assign this group to those objects to allow only users within the group to run them.

3. Create a System View Node

Now we will create a *System View* node to hold this example's data, analyses, and report configuration. In the tree view, click the plus sign ⊞ adjacent to the *System View* node to expand it. Right-click on the *STATISTICA Enterprise* folder, and from the shortcut menu, select ***New Folder***. In the *Folder name* text box in the properties page, enter *Test Example 1* as the new folder's name.

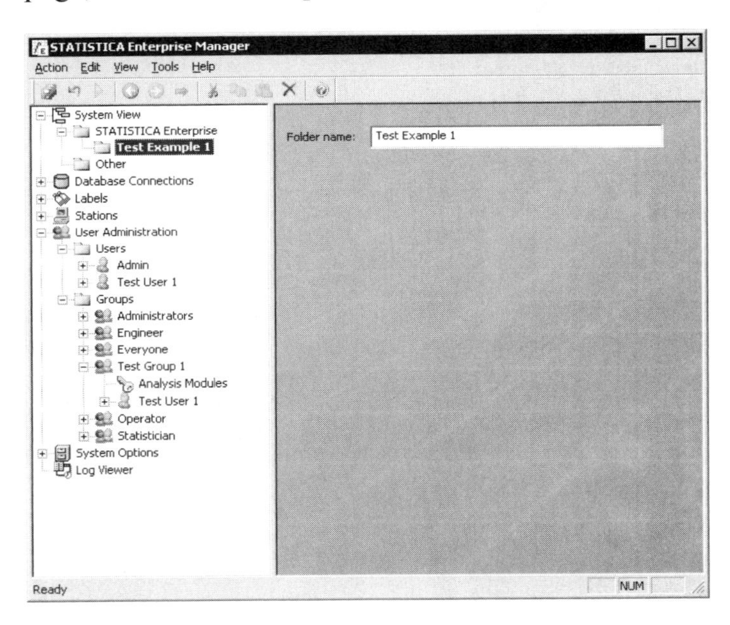

Click the ***Commit Changes*** icon on the toolbar to commit the change. This folder will now be used to house the data, analysis, and report configurations.

4. Create a New Database Connection

Right-click on the *Database Connections* node in the tree view, and from the shortcut menu, select ***New Database Connection*** to display the ***Data Link Properties*** dialog.

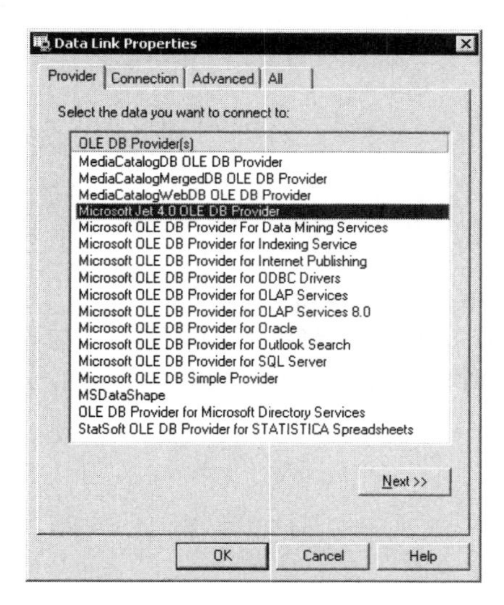

Select ***Microsoft Jet 4.0 OLE DB Provider*** and click the ***Next*** button (if this provider is not available to you, please visit this Web site – msdn2.microsoft.com/en-us/data/aa937712.aspx#MDAC – to install the updated MDAC or Jet provider before proceeding, and then restart this step). On the ***Connection*** tab, click the button adjacent to the ***Select or enter a database name*** field, and browse to the location to which *STATISTICA* has been installed (by default *C:\Program Files\StatSoft\STATISTICA*). Double-click the *Examples* folder and then the *Database* folder, and select the *ProcessData.mdb* file and click the ***Open*** button.

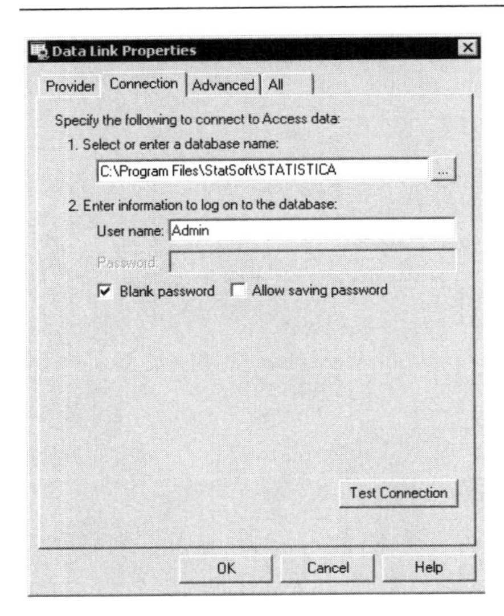

On the ***Connection*** tab, click the ***Test Connection*** button. A prompt will be displayed that acknowledges that the *Test connection succeeded* (if it didn't succeed, check your access permissions to the file and check your MDAC installation).

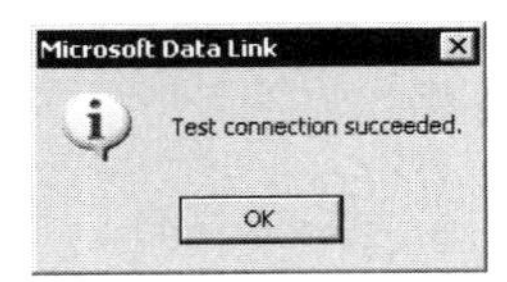

Click ***OK*** in the prompt, and click ***OK*** in the ***Data Link Properties*** dialog. In the resulting property page, enter *Test Example Connection 1* in the ***Name*** field.

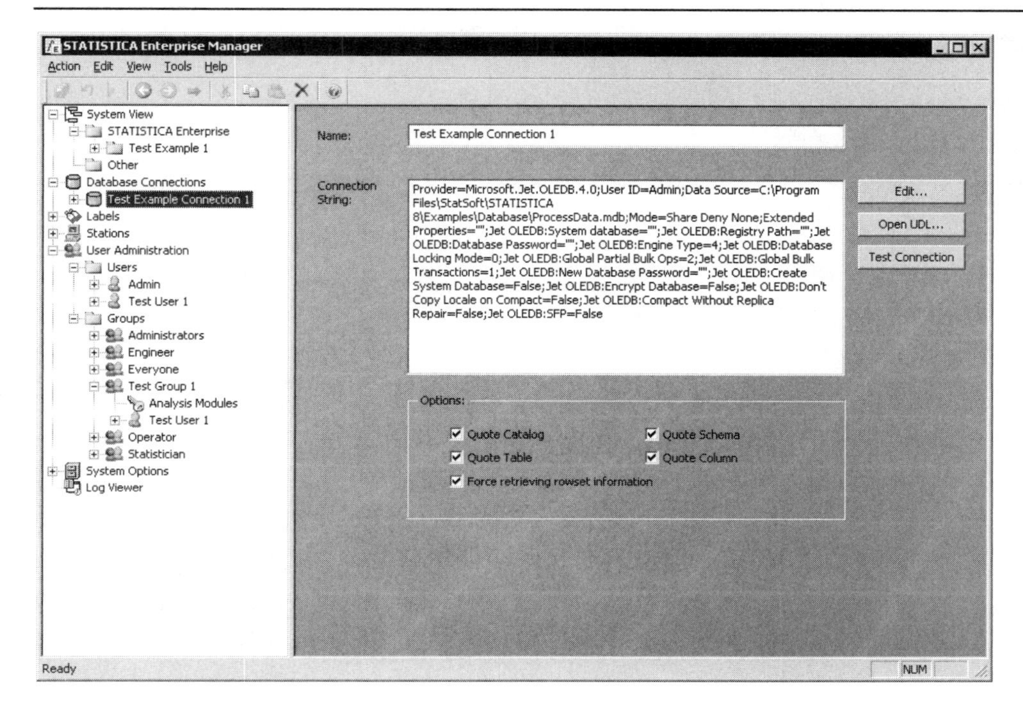

Now, click the ***Commit Changes*** toolbar button.

With the Database Connection created to the *ProcessData* database, we will now create a data configuration to extract data from the database.

5. Create a Data Configuration

Right-click on the *Test Example 1* folder in the tree view, and from the shortcut menu, select ***New Data Configuration***. In the property page, enter *Test Example 1* in the ***Name*** field. Click the drop-down arrow next to the ***Connection*** field, and from the drop-down list, select *Test Example Connection 1*.

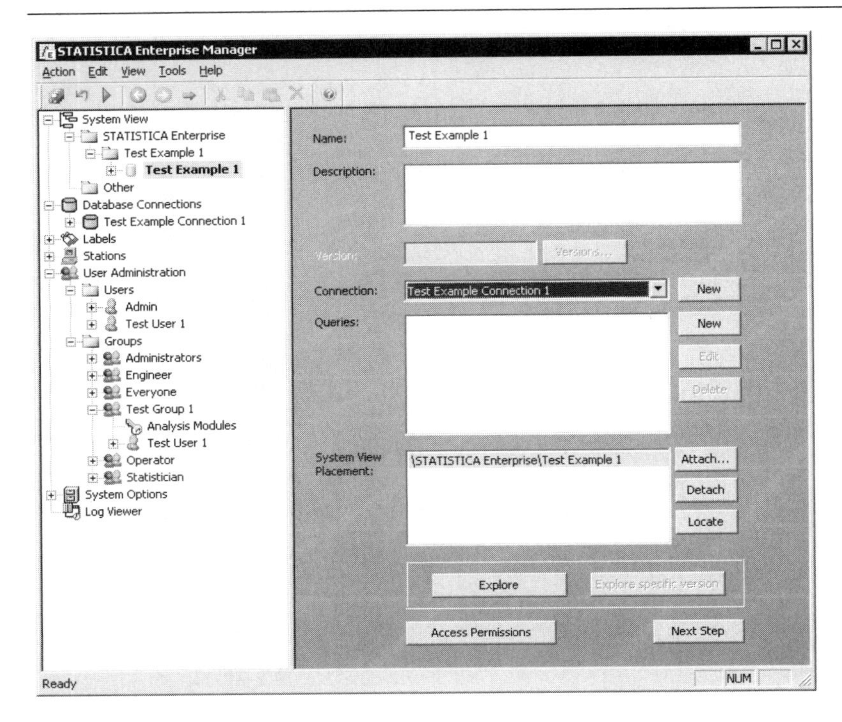

Click the ***Next Step*** button in the lower-right of the property page to display the new query options.

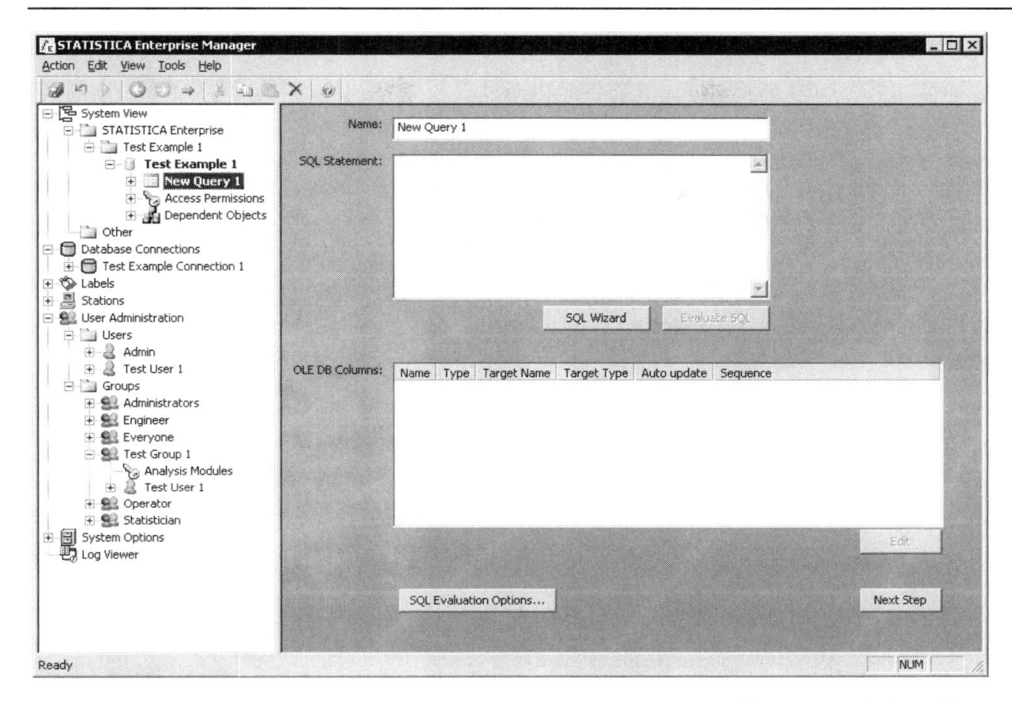

Click the ***SQL Wizard*** button to display the ***New Query*** dialog, which will open in *STATISTICA*.

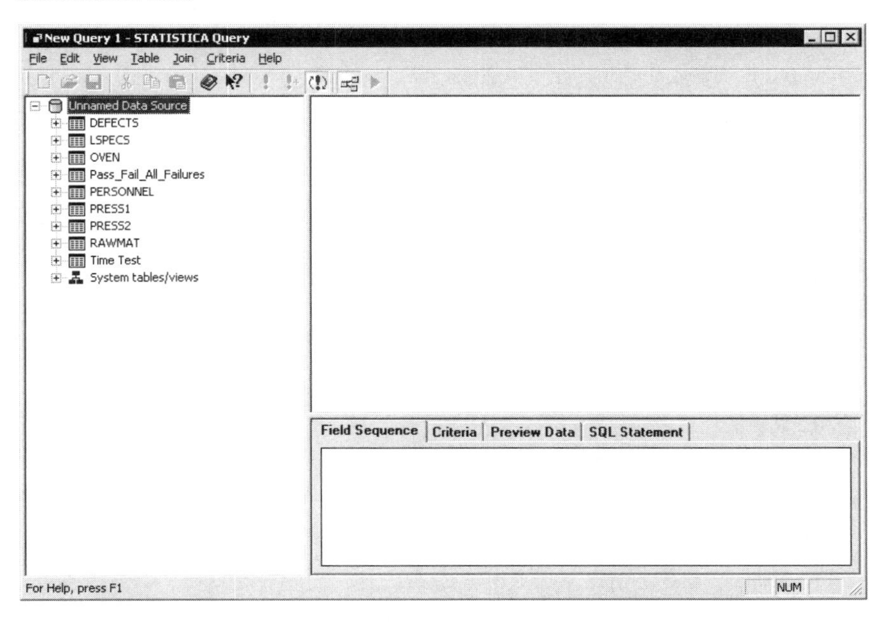

Drag the *RAWMAT* table from the left pane into the editor viewer (the upper-right pane), and then select, in the following order, the *ID*, *OPERATOR*, *LOCATION*, and *HEIGHT* fields.

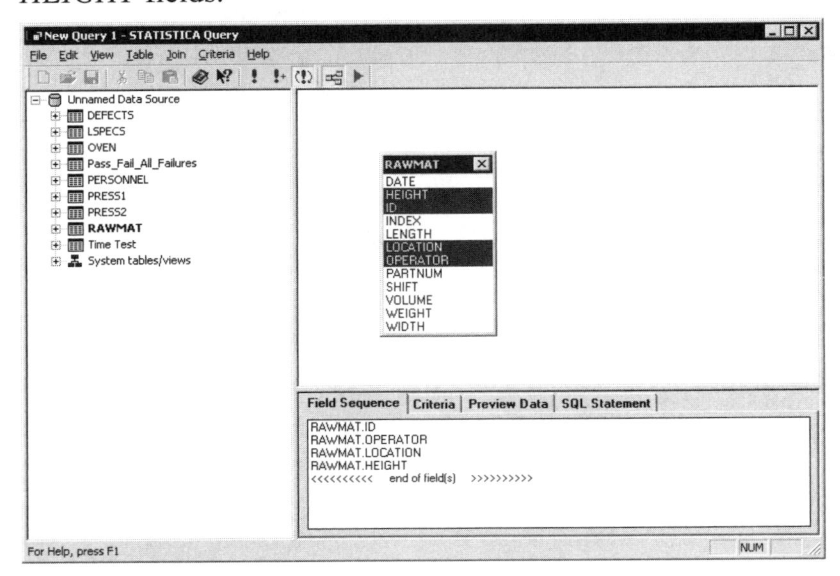

Select the ***Preview Data*** tab in the query properties view (lower-right pane) and click the ***Refresh*** **!** toolbar button (the red exclamation mark). This will test the query to ensure that values are being retrieved from the defined query.

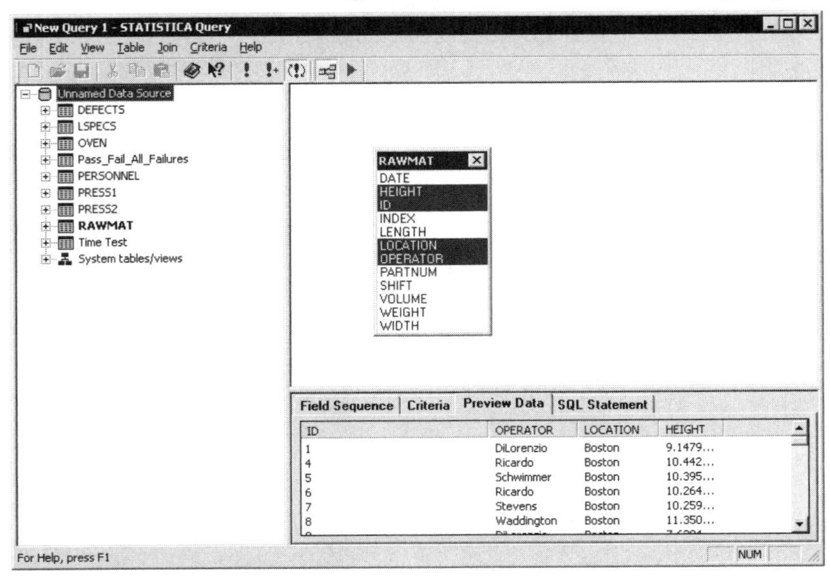

Click the ***Return Data to STATISTICA*** toolbar button (green arrow) to submit this query back to the data configuration. When prompted to evaluate the SQL, click the ***Yes*** button, and the ***OLE DB Columns*** will be resolved.

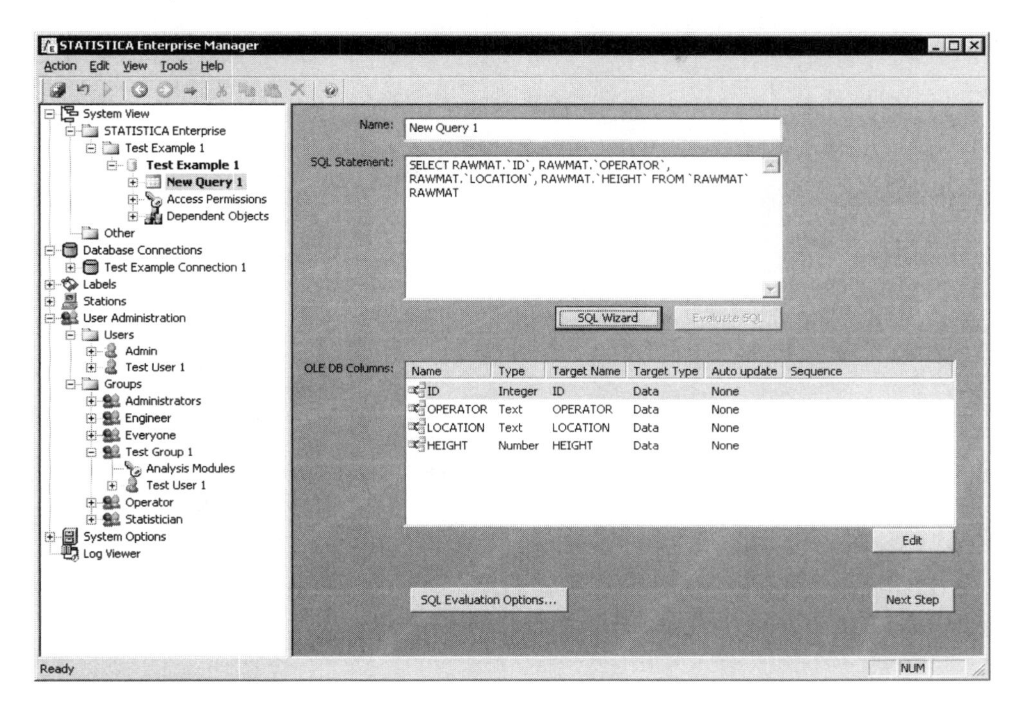

With the *ID* row highlighted, click the ***Edit*** button to display options to edit the *ID* column. Click the drop-down arrow next to the ***Auto Update*** field, and from the drop-down list, select ***First update column***. This allows you to detect changes in the ID column. In addition, the column is sorted.

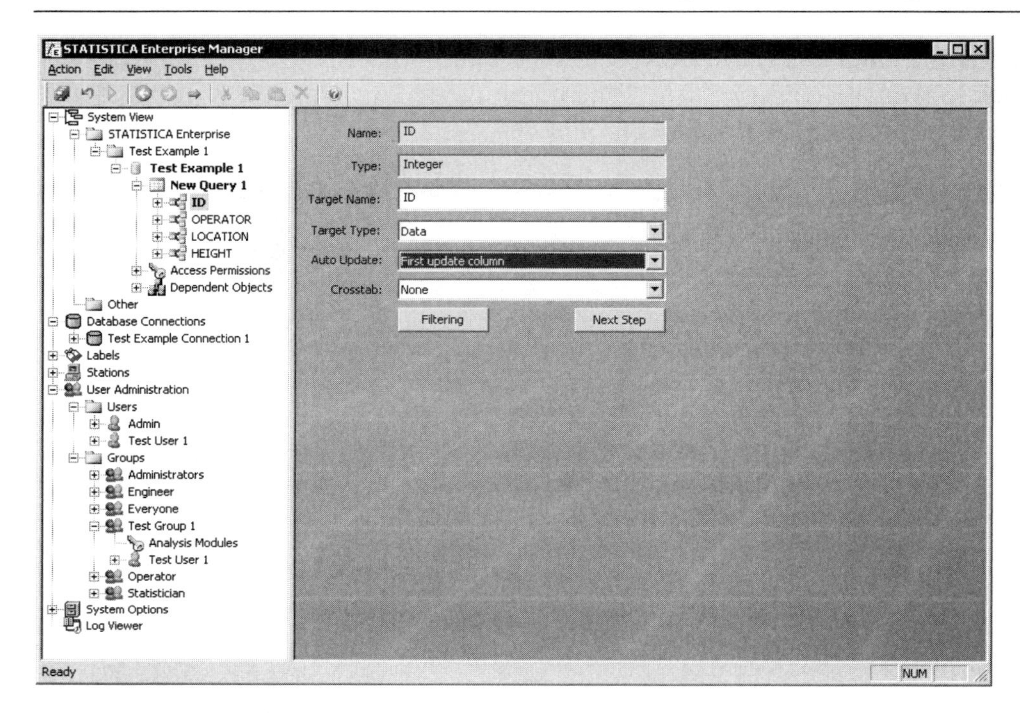

Click the ***Next Step*** button to edit the *OPERATOR* column. Click the ***Filtering*** button to display the filtering options, and select the ***Enabled*** check box to allow filtering on the *OPERATOR* column.

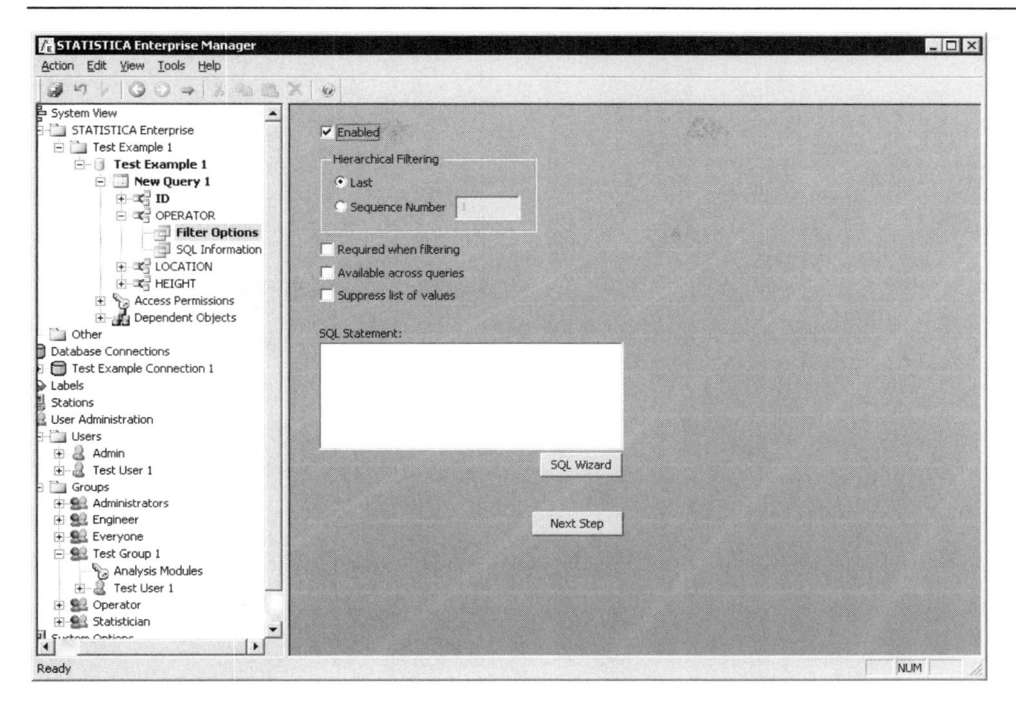

Click the ***Next Step*** button to return to *OPERATOR* column editing options, and then click the ***Next Step*** button to edit the *LOCATION* column. Click the ***Filtering*** button to display the filtering options, and select the ***Enabled*** check box to allow filtering on the *LOCATION* column. Click the ***Next Step*** button to return to the *LOCATION* column editing options, and then click the ***Next Step*** button to edit the *HEIGHT* column. Click the drop-down arrow next to the ***Target Type*** field, and from the drop-down list, select ***Variable Characteristic.* This** will make this column available to perform packaged SPC analyses (this is the column containing the data to be analyzed).

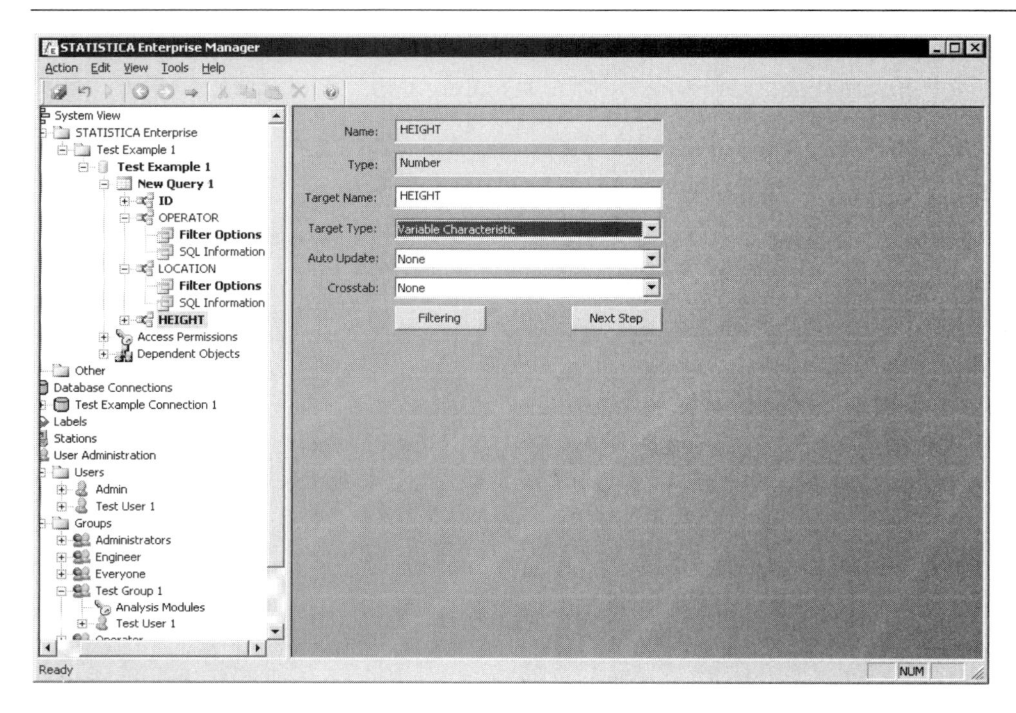

Next, click the ***Next Step*** button to display the ***Access Permissions*** options for this object. From the list of ***Available Users and Groups***, select *Test Group1*, and then click the top arrow button [→] to move *Test Group 1* to the ***Access Permissions*** list. This will make this data configuration executable (but not editable) by the users of *Test Group 1*.

Click the ***Commit Changes*** toolbar button to commit this new data configuration to *STATISTICA Enterprise Manager*.

6. Create an Analysis Configuration

Now that a data configuration has been defined to extract data from the *ProcessData.mdb* database, an analysis configuration to analyze the data will need to be created.

In the tree view, right-click on the *Test Example1* folder, and from the shortcut menu, select ***New Analysis Configuration*** to display the ***Select a Data Configuration*** dialog. Select the *Test Example 1* object and click the ***OK*** button. If a dialog is

displayed with the statement, "When selected, this option will replace the permissions of this analysis with those of the selected data," click ***OK***.

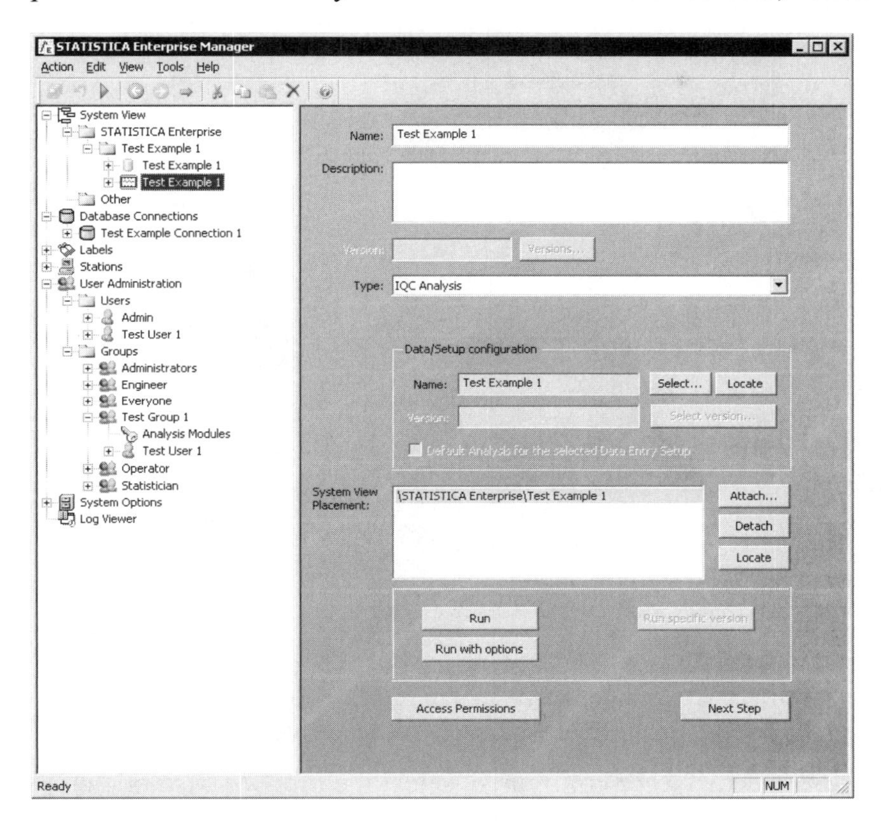

Click the ***Next Step*** button to continue creating the analysis configuration (leaving the default name the same as the data configuration for expediency only). Click the ***Next Step*** button once again to continue editing the analysis configuration.

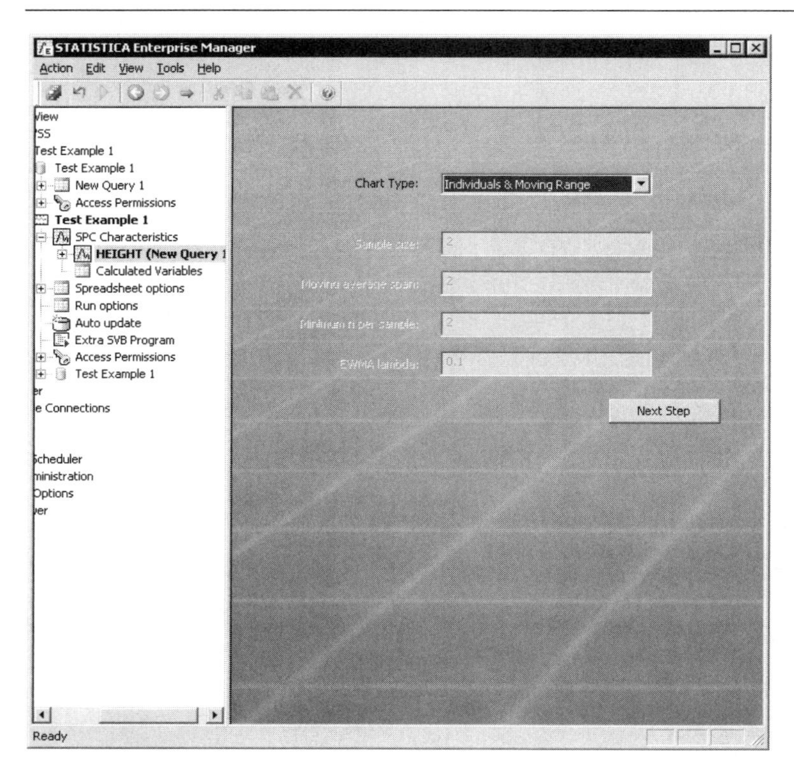

In the property page for the *SPC Characteristics - HEIGHT* column, change the ***Chart Type*** to ***Individuals & Moving Range*** (as shown in the above illustration).

Since no other SPC options need to be configured, select the *Run options* node in the tree view, and select the ***Show SQL Criteria dialog*** check box in the property page.

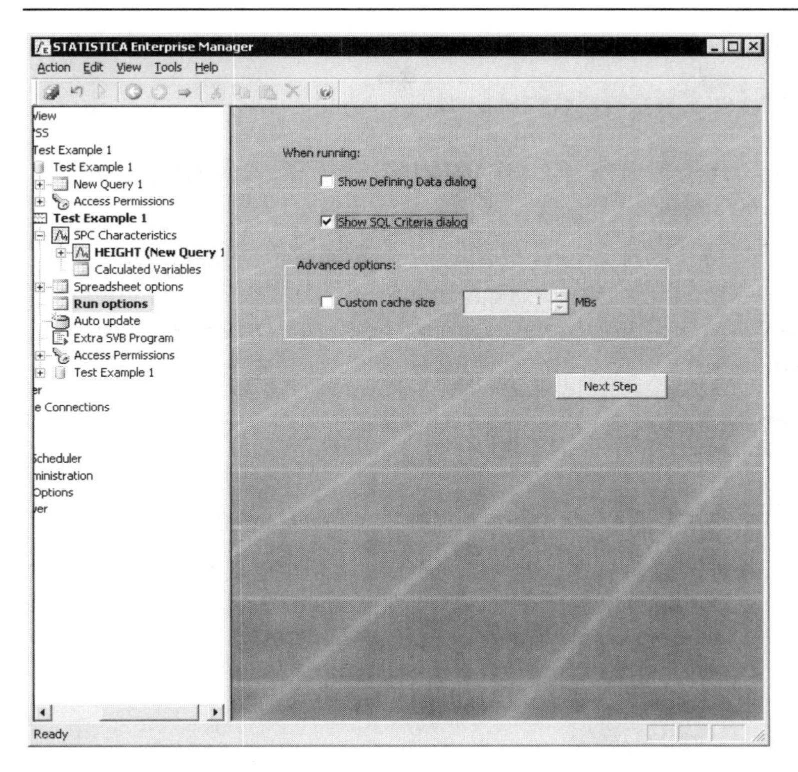

This option will specify that *STATISTICA* prompt for filtering on those columns that have ***Filter*** options in the data configuration (if, when defining the ***Filter*** options, they were set to ***Required when filtering***, this step would not be required as it would always force a filtering prompt when running – in this example it was not required to force filtering).

Click the ***Commit Changes*** toolbar button to commit this analysis configuration to *STATISTICA Enterprise*.

7. Run the Analysis Configuration

Close the *Enterprise Manager*, and log on to *STATISTICA* as the *Test User 1* user created in Step 1.

From the ***Enterprise*** menu, select ***Run Analysis/Report*** to display the ***Run Analysis or Report*** dialog (this dialog may be displayed automatically depending on your configuration). Select the *Test Example 1* analysis, and click the ***OK*** button; the ***SQL Criteria*** dialog will be displayed.

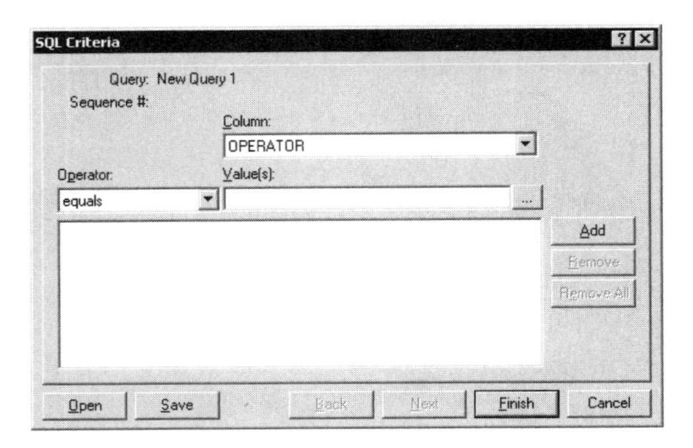

Click the drop-down arrow next to the ***Column*** field and select *LOCATION* from the drop-down list. Click the browse button [...] to display the ***Value of LOCATION*** dialog, which contains the list of available *LOCATION* values. Select *Boston* and click the ***OK*** button.

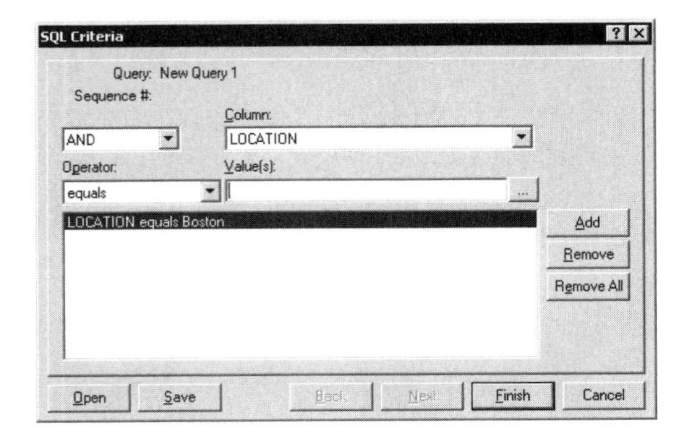

Click the ***Finish*** button to complete the filtering step, extract the data, and perform a packaged analysis on the *HEIGHT* column.

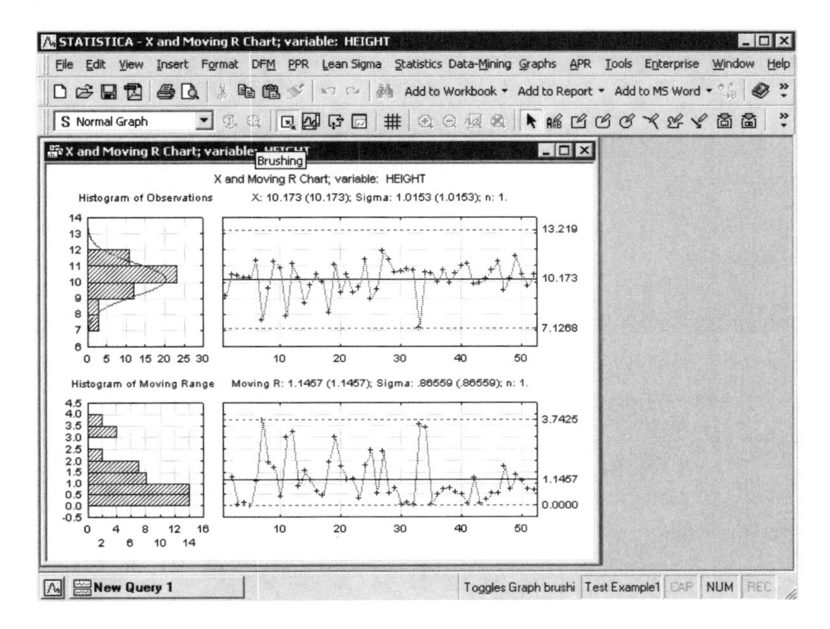

Custom User Interfaces

Note that this simple example illustrates how to enable and run an analysis configuration using the standard *STATISTICA* user interface and output components.

However, one of the major strengths of *STATISTICA Enterprise* is the ease of creating custom user interfaces (e.g., for different categories of users depending on their roles in the organization, expertise, or data access privileges).

You can easily create a customized user interface at any degree of complexity, from highly simplified ones, e.g., one that contains only three options:

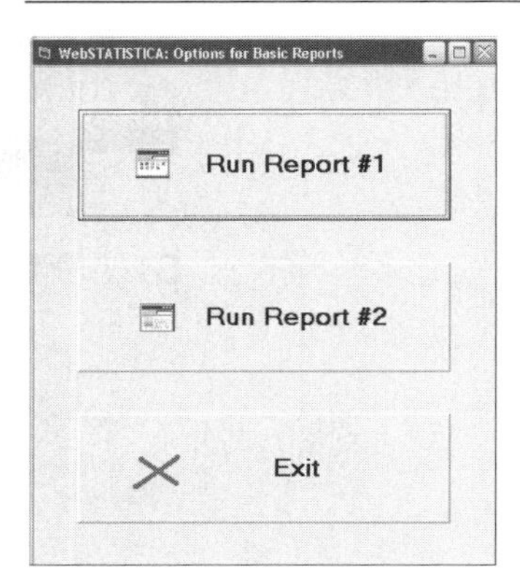

to very elaborate user interfaces of virtually unlimited flexibility:

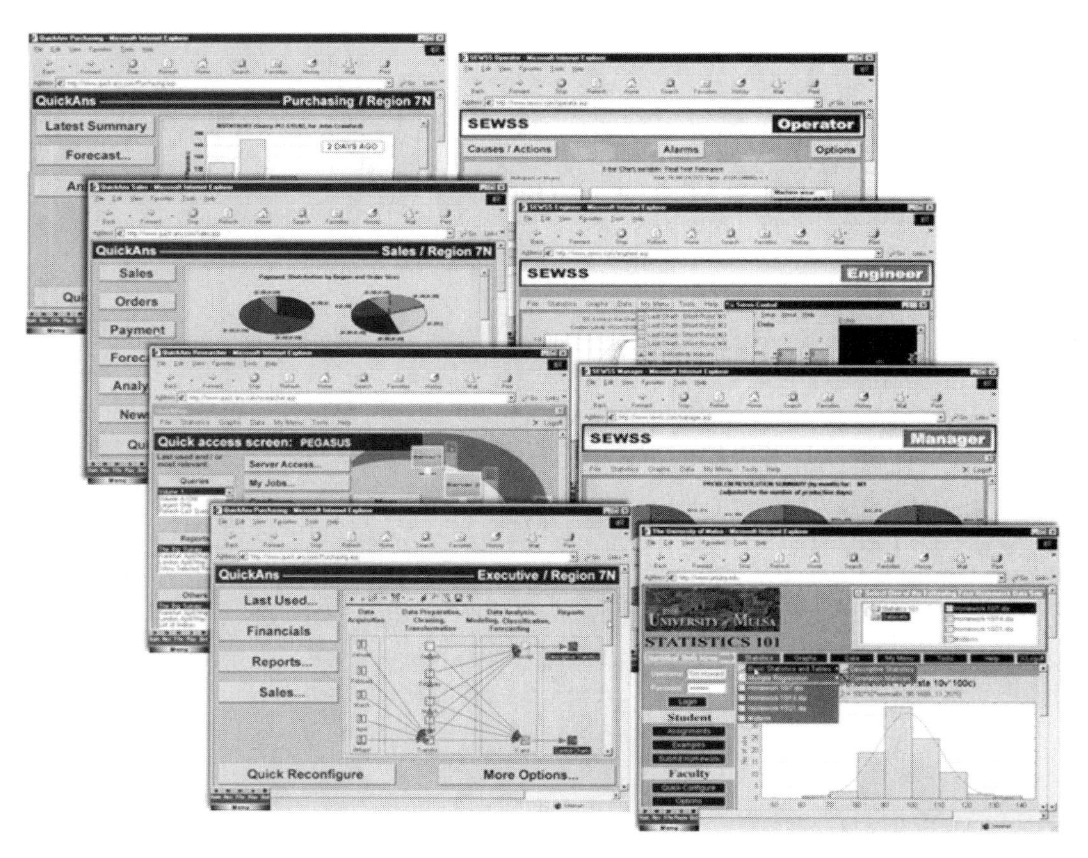

Please refer to the *STATISTICA Enterprise* documentation for details and examples.

CHAPTER

USER INTERFACE

USER INTERFACE

GENERAL FEATURES

Customized Operation

The *STATISTICA* system can be controlled in several ways. The following sections summarize the features of the main alternative user interfaces of *STATISTICA*:

1. Interactive interface (see page 132)
2. *STATISTICA* Visual Basic (see page 144)
3. Web browser-based interfaces (see page 145)
4. Microsoft Office Integration (see page 146)

However, note that:

- Many aspects of these user interfaces do not exclude each other; thus, depending on your specific applications and preferences, you can combine them;
- The customizable menus and toolbars can be used to integrate the alternative user interfaces and, for example, to provide quick access to macro (Visual Basic) programs or commonly used files; and
- Almost all features of these alternative user interfaces can be customized (leading to a different appearance and behavior of *STATISTICA*); it is generally recommended that you customize your system in order to take full advantage of *STATISTICA*'s potential to meet your preferences and optimal requirements of

the tasks that you need to accomplish (see *Customization of the Interactive User Interface* on page 221).

Alternative Access to the Same Facilities; Custom Styles of Work

Even without any customization, the default settings of *STATISTICA* offer alternative user interface means and solutions to achieve the same results. This "alternative access" principle present in every aspect of its user interface enables *STATISTICA* to support different styles of work. For example, most of the commonly used tools can be accessed alternatively:

- From traditional menus
- Via keyboard shortcuts
- By using the toolbars and the clickable fields on the status bar
- Via custom toolbars (user-defined toolbars with buttons and special controls, which can include macros and commands)
- From the shortcut menus associated with specific objects (cells, workbook icons, parts of graphs), which are displayed by right-clicking on the item.

It is suggested that you explore the alternative user interface facilities of *STATISTICA* before becoming attached to one style or another.

MULTIPLE ANALYSIS SUPPORT

As mentioned before, you can have several instances of *STATISTICA* open at the same time. Each of them can run the same or different types of analyses (traditionally called modules), such as ***Basic Statistics***, ***Multiple Regression***, ***ANOVA***, etc. Moreover, in one *STATISTICA* instance, multiple analyses can be open simultaneously. They can be of the same or a different kind (e.g., five ***Multiple Regressions*** and two ***ANOVAs***), and each of them can be performed on the same or a different input data file (multiple input data files can be opened simultaneously).

Individual "analyses" – functional units of your work. In order to facilitate taking advantage of this "multitasking" functionality, your work with *STATISTICA* is organized into functional units called "analyses" that are represented with buttons on the analysis bar at the bottom of the application window (above the status bar, see the following illustration, where ***Basic Statistics***, ***Cluster Analysis***, and ***Canonical Analysis*** are running simultaneously). Consecutive buttons are added as you start new analyses. A variety of options is provided to control (and/or permanently configure) this aspect of *STATISTICA*.

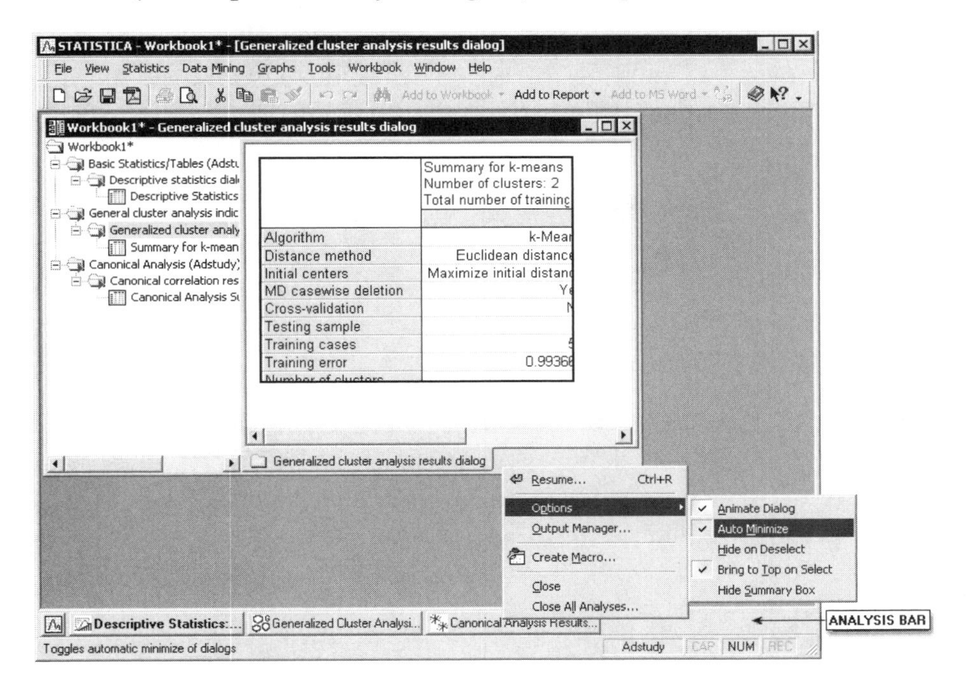

By default, when you select specific output from a results dialog, the output (a table or a graph) is displayed and the dialog is automatically minimized into its respective analysis button on the bottom of the screen. Click that button (or press CTRL+R) to display the dialog again and resume the analysis.

A selection of options pertaining to analysis management is available on the shortcut menu (accessed by right-clicking on an analysis button on the analysis bar) related to each respective analysis button (as shown above).

A useful hint for those with large screens. If you have a large screen, you can turn off the default minimization of the analysis dialogs and take advantage

of the fact that most of these dialogs are small and, thus, can remain on the workspace without interfering with the viewing of analysis results. You can adjust this option either for a particular analysis (clear the ***Auto Minimize*** command on the analysis button shortcut menu, shown above), or globally for the entire program (use the ***Analyses/Graphs*** tab of the ***Options*** dialog, accessible via the ***Tools - Options*** menu).

When you run multiple analyses and the *STATISTICA* workspace becomes cluttered, you can hide all windows related to specific analyses (or close them altogether via the analysis button shortcut menu command ***Close All Analyses***). You can also open new *STATISTICA* instances, which offers another simple way to organize and manage your work.

INTERACTIVE USER INTERFACE

Overview

Main components of the interactive user interface of *STATISTICA*. Although the interactive user interface of *STATISTICA* is not the only one available (see Chapter 8 – *Customizing STATISTICA,* page 221 and Chapter 9 – *STATISTICA Visual Basic*, page 227), in most cases it is the easiest and most commonly used. Many components of this user interface can be seen in the *STATISTICA* application window.

First, similar to most software programs, menu bars and various toolbars are displayed at the top of the window. These are customizable and displayed in the most appropriate manner for your tasks.

At the bottom of the window, the analysis bar (containing minimized analysis/graph dialogs) and the status bar are displayed. Additionally, shortcut menus are available when you right-click in appropriate places.

Data files can be displayed in spreadsheets, workbooks, reports, or individual windows. Results spreadsheets or graphs can be displayed in workbooks, reports, or individual windows. Note that additional documents (such as Word or Bitmap images) can also be displayed in spreadsheets, workbooks, or reports. Finally, *STATISTICA* Visual Basic code is displayed in macro windows.

Normally you would not simultaneously see all of these facilities and tools at one time. You always have the ability to make the user interface of *STATISTICA* as simple or complex as your particular needs and comfort level demand (see page 221). These various tools and facilities are described in detail in the *Electronic Manual*.

Modules. While *STATISTICA* offers a variety of statistical and graphical procedures, each procedure can be performed in the same instance of *STATISTICA*. This means that, for example, it is possible to calculate residual statistics using options in the ***Multiple Regression*** module, then immediately use that output in the ***Factor Analysis*** or another exploratory module without first starting another instance of *STATISTICA*. For more information on using results as input data, see *Can I Use the Results of My Analysis to Perform Another Analysis?* in the *Electronic Manual*.

The Flow of Interactive Analysis

Startup Panel. When a statistical procedure is selected from the ***Statistics***, ***Data-Mining***, or ***Graphs*** menu, its respective Startup Panel is displayed (as shown below; ***Basic Statistics/Tables*** was selected from the ***Statistics*** menu to display the ***Basic Statistics and Tables*** Startup Panel).

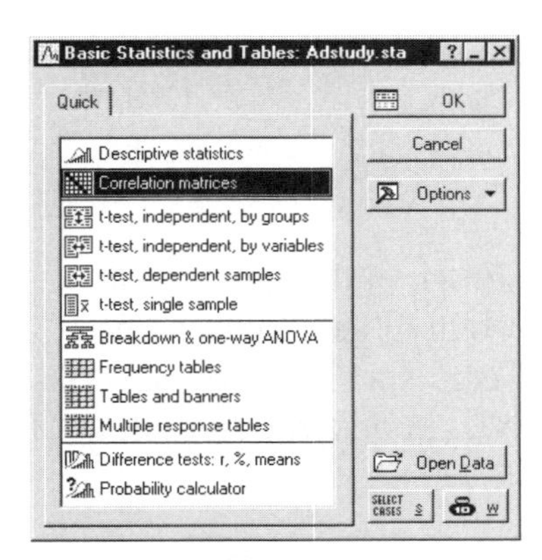

Each Startup Panel contains a list of the types of analyses available in that particular module. Clicking anywhere outside the panel automatically minimizes it as a button on the analysis bar. If your system includes a high-resolution screen, you can change this default and keep the consecutive dialogs (in each analysis sequence) on the workspace.

Toolbars. If you prefer to use buttons rather than menus to select statistical analyses, you can activate the ***Statistics*** toolbar (which contains buttons for every module) by right-clicking on any toolbar and selecting ***Statistics*** from the shortcut menu that lists available toolbars. Alternatively, you can select ***Statistics*** from the ***View - Toolbars*** menu. You can also create your own toolbar that holds the toolbar buttons for analyses you use most frequently (as illustrated on page 143). For more information on toolbars, see the *Toolbar Overview* in the *Electronic Manual*.

Analysis specification and output selection (results) dialogs. When the desired analysis is selected on the Startup Panel, the analysis specification dialog is displayed, in which you select the variables to be analyzed and other options and features of the task to be performed. Often, these dialogs contain several tabs that group the options, analyses, and/or results in logical categories to make it easier to locate specific features.

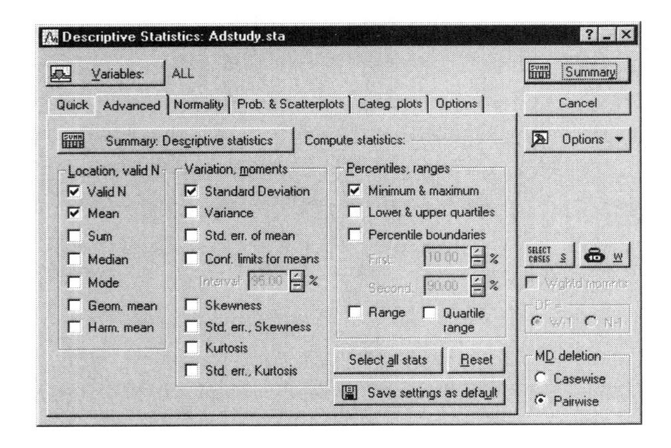

In some simple analyses (such as ***Descriptive Statistics***, shown in the illustration above), the analysis specification dialog also serves as an output selection dialog where you can specify the type and format of the output (e.g., specific spreadsheets or graphs). Most analyses, however, have a separate analysis specification dialog and results dialog.

Spreadsheet facilities for scenario (what-if) analyses and customized appearance. *STATISTICA* provides you with the capability to append supplementary information about variable measurement types and case states to your spreadsheets. This "metadata" can be used to create a more comprehensive description of your data set, facilitate "what-if" types of exploratory analyses, and customize the appearance of cases in graphs.

Case states and brushing. You can assign case states to cases in order to customize the appearance of points in graphical displays, thus making it very easy to identify influential and interesting points. A wide selection of symbols and colors is available to customize the appearance of selected points. Not only can case states be assigned in the spreadsheet before a graph is made, they can also be assigned interactively in the graph via the ***Brushing*** facilities (accessible by selecting ***Brushing*** from the ***View*** menu when a graph is open). The case states assigned in the graph propagate back to the spreadsheet. The ability to assign case states in either the spreadsheet or graph further facilitates the exploratory visual analysis of data.

Measurement types and automatic variable pre-screening. The modeling or measurement type of a variable can be explicitly defined in order to indicate what analyses and graphs are appropriate for such a variable. These measurement types will map directly to subsequent analyses and graphs, identifying appropriate variables in each case (e.g., variables of type categorical will be present within the list of categorical predictors available in a *Factorial ANOVA*).

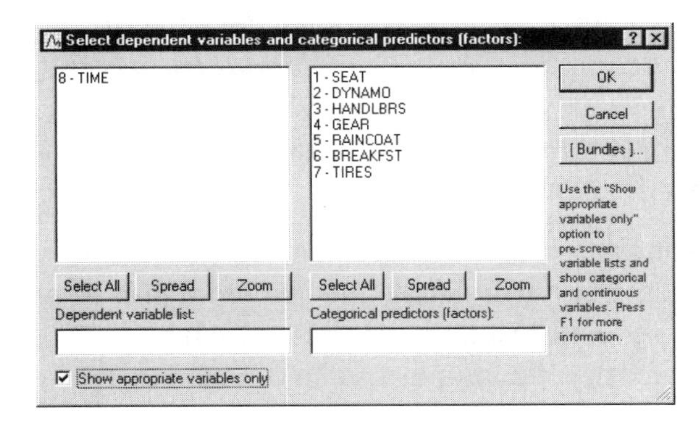

On all variable selection dialogs (such as the one shown above), the ***Show appropriate variables only*** option is provided, which enables you to pre-screen or filter variables according to their ***Measurement Type*** (specified in the ***Variable*** specification dialog, accessible by double-clicking on a variable header); if that type is *Auto*, then the ***Automatic variable pre-screening and classification*** options (specified on the ***Analysis/Graph*** tab of the ***Options*** dialog, accessible by selecting ***Options*** from the ***Tools*** menu) determines how *STATISTICA* will automatically determine the ***Measurement Type***.

Auto filtering (cloaking variables and cases). Filtering (accessible from the ***Data - Auto Filtering*** submenu) is a quick and easy way to display a specific portion of the data in your spreadsheet without sorting the data or creating a subset. When a variable is filtered, only the values that meet the specified criteria are displayed in the spreadsheet. Cases that do not meet the criteria are hidden from sight but not removed from the spreadsheet (e.g., in the spreadsheet shown below, only the cases for *GENDER* = *MALE* are displayed).

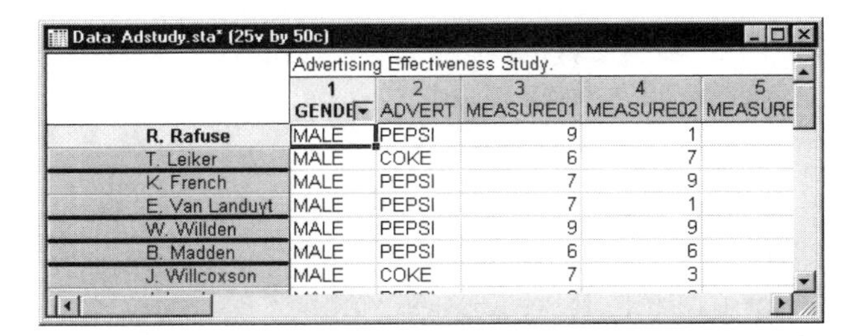

Data: Adstudy.sta* (25v by 50c)

Advertising Effectiveness Study.

	1 GENDE	2 ADVERT	3 MEASURE01	4 MEASURE02	5 MEASURE
R. Rafuse	MALE	PEPSI	9	1	
T. Leiker	MALE	COKE	6	7	
K. French	MALE	PEPSI	7	9	
E. Van Landuyt	MALE	PEPSI	7	1	
W. Willden	MALE	PEPSI	9	9	
B. Madden	MALE	PEPSI	6	6	
J. Willcoxson	MALE	COKE	7	3	

Although they are hidden, they are still available for statistical and graphical analyses.

Output. As described in more detail in Chapter 5 – *Five Channels for Output From Analyses* (page 151) and as illustrated in *Example 1: Correlations* (page 11) and *Example 2: ANOVA* (page 34), the consecutive output spreadsheets and graphs are displayed in workbooks by default. These workbooks can be saved and later reopened, making it easy to return to specific results as needed.

Additionally, you can send all output to an analysis report (see page 154), which produces an easily organized (via the report tree), formatted, and printed report of a specific analysis. You can also choose to send all results, regardless of what analysis it comes from, to a single report. Lastly, the output can be directed to separate windows.

To specify output options for a single analysis or session, click the Options button on the analysis or graph specification dialog and select ***Output*** to display the ***Analysis/Graph Output Manager*** dialog.

Global output options are available by selecting ***Options*** from the ***Tools*** menu and accessing the ***Output Manager*** tab of the ***Options*** dialog or by selecting ***Output Manager*** from the ***File*** menu. For more information, see the *Electronic Manual.*

Features of Analyses

STATISTICA provides direct access to all statistical analysis dialogs via the ***Statistics*** menu:

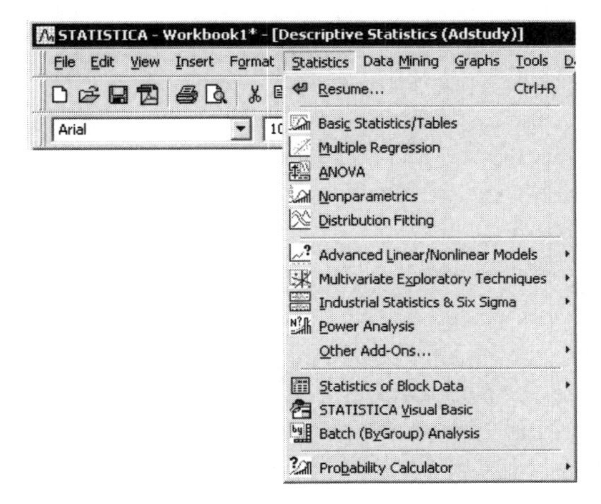

and the ***Data Mining*** menu:

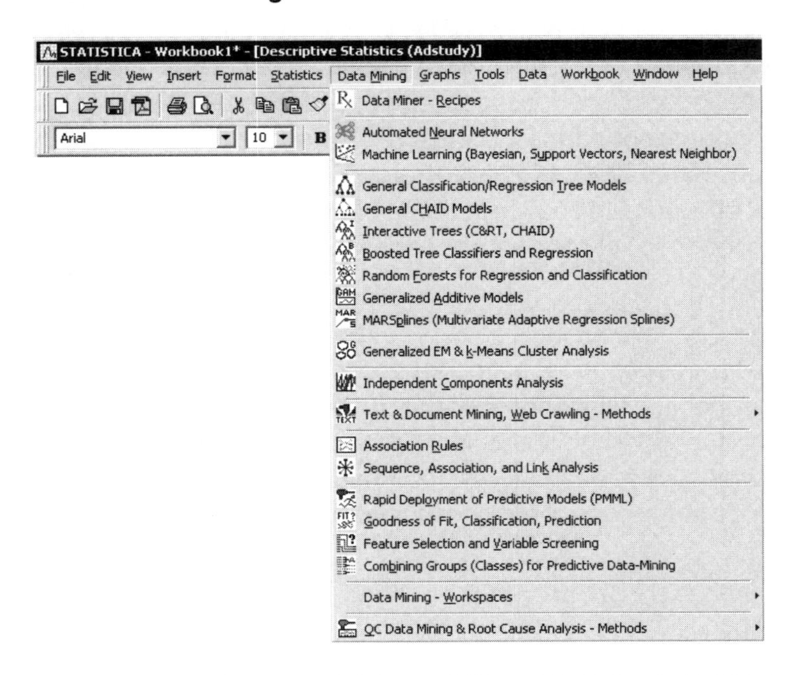

and provides direct access to all graphical analysis dialogs via the ***Graphs*** menu:

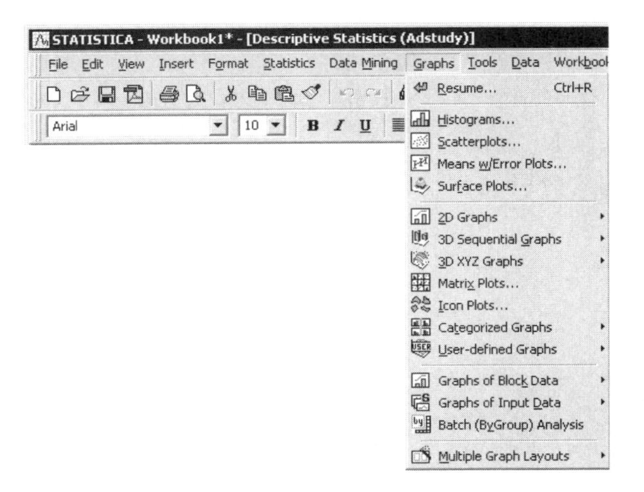

(as well as from the ***Statistics***, ***Data Mining***, and ***Graphs*** toolbars). These menus are never disabled, i.e., they are available whenever any input data document is open. The ***Statistics*** and ***Data Mining*** menus provide access to all available analysis types within *STATISTICA*. The ***Graphs*** menu provides direct access to a variety of commonly used graph types (e.g., scatterplots, histograms, means/error plots, etc.) as well as hierarchical access to all graph types in *STATISTICA* including ***2D Graphs***, ***3D Sequential*** and ***XYZ Graphs***, ***Matrix Plots***, ***Icon Plots***, ***Categorized Graphs***, ***User-defined Graphs***, ***Graphs of Block Data***, ***Graphs of Input Data***, and ***Multiple Graph Layouts***. Comprehensive discussions of all the various types of statistics and graphs offered by *STATISTICA* are available in the glossary of the *Electronic Manual*. See also, *Appendix B: STATISTICA Family of Products* (page 281) for more information on all members of the comprehensive selection of data analysis applications from the *STATISTICA* family of products.

Using the analysis bar. To take advantage of *STATISTICA*'s "multitasking" functionality (see *Multiple Analysis Support*, page 130), *STATISTICA* analyses are organized as functional units that are represented with buttons on the analysis bar at the bottom of the application window (above the status bar, see the illustration below, where ***Basic Statistics***, ***Cluster Analysis***, and ***Canonical Analysis*** are running simultaneously). Consecutive buttons are added as you start new analyses.

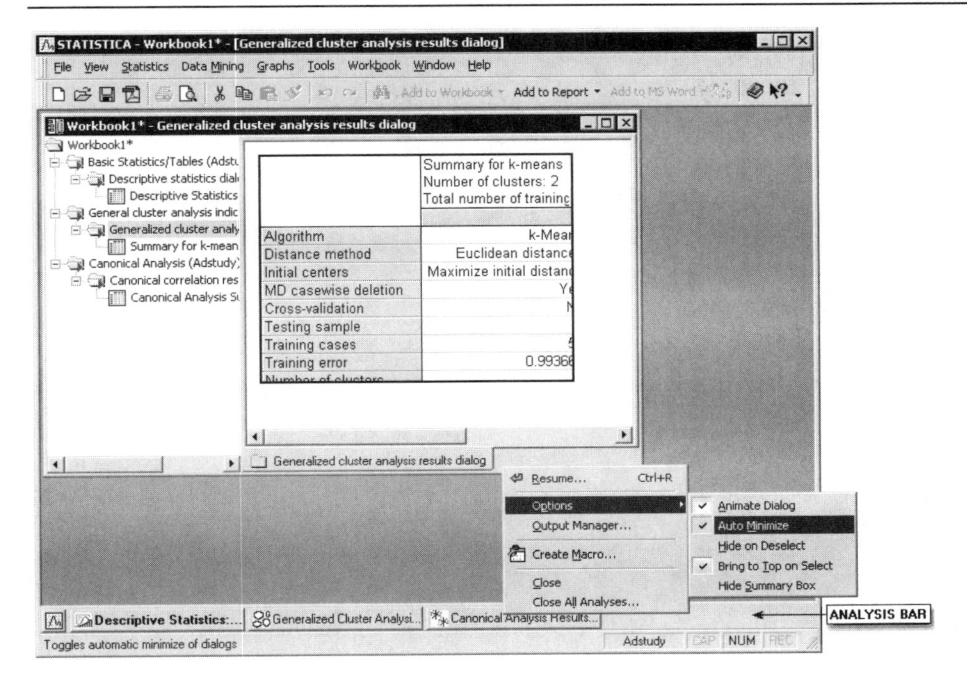

Minimizing dialogs (and a hint for users with large screens). Depending on your preferences, you can choose to minimize all analysis dialogs when you select another window in *STATISTICA* or another application. By default the ***Auto Minimize*** command is selected; however, when your screen is large enough to accommodate several windows, it is recommended that you clear this command. This keeps the analysis dialogs on screen while the respective output created from these dialogs is produced, thus enabling you to use the dialogs as "toolbars" from which output can be selected. See page 131 for information on how to adjust this command.

Continuing analyses/graphs. It is easy to continue the current analysis or graph (i.e., to change the focus to the current dialog for a particular analysis). Select ***Resume*** from the ***Tools - Analysis Bar*** submenu, press CTRL+R, or click the analysis/graph button on the analysis bar. When multiple analyses are running, you can also select the specific analysis from the ***Tools - Analysis Bar - Select Analysis/Graph*** submenu (as shown below).

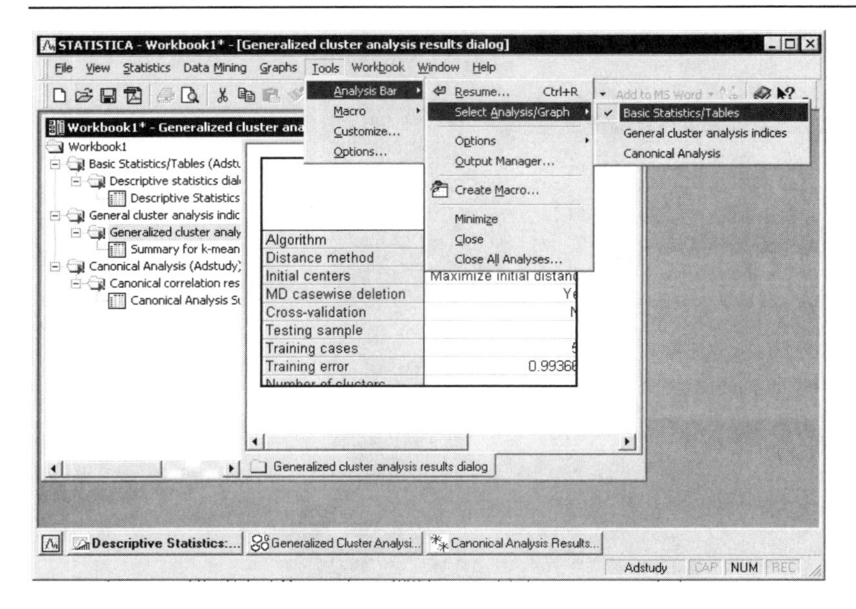

Hiding windows. To further facilitate the organization of windows from various analyses, you can hide all windows associated with a particular analysis when that analysis is deselected by selecting ***Hide on Deselect*** from the ***Tools - Analysis Bar - Options*** submenu. By default, this command is not selected. Note that this command only applies when the results are sent to individual windows; see the discussion of the *Output Manager* (page 151) for more details on managing output from analyses. In addition, there is a command to close all document windows, ***Window - Close All*** (or CTRL+L), and a command to close all analyses, ***Tools - Analysis Bar - Close All Analyses***.

Bringing windows to the top. Select ***Bring to Top on Select*** from the ***Tools - Analysis Bar - Options*** submenu to activate (bring to the top of *STATISTICA*) all windows associated with a particular analysis when that analysis is selected, replacing whatever dialogs were on top. This command also facilitates the organization of individual windows from various analyses. By default, this command is selected. Note that this command only applies when the results are sent to individual windows; see the discussion of the *Output Manager* (page 151) for more details on managing output from analyses.

Hiding the summary box. By default, a summary box is located at the top of certain results dialogs (such as ***Multiple Regression Results***) and contains basic summary information about the analysis. You can hide an individual summary box by clicking the button in the lower-right corner of the summary box. You can also suppress the display of all summary boxes globally by selecting ***Hide Summary Box*** from the ***Tools - Analysis Bar - Options*** submenu.

Document Types

STATISTICA uses seven principal document types:

- Workbooks (see pages 152 and 177)
- Spreadsheets (multimedia tables) (see page 181)
- Reports (see pages 154 and 185)
- Graphs (see pages 188 and 197)
- Macros (*STATISTICA* Visual Basic programs) (see pages 189 and 227)
- *STATISTICA* Project Files (see page 190)
- Data Miner Project Files (see page 193)

Using these seven document types, you can manage data of various types, perform data entry and analysis, generate graphs of the highest quality, develop custom applications of any degree of complexity, and create custom-formatted reports.

You can quickly access the most recently used documents. Click the *STATISTICA* Start button (in the lower-left corner of the screen) and select ***Documents***.

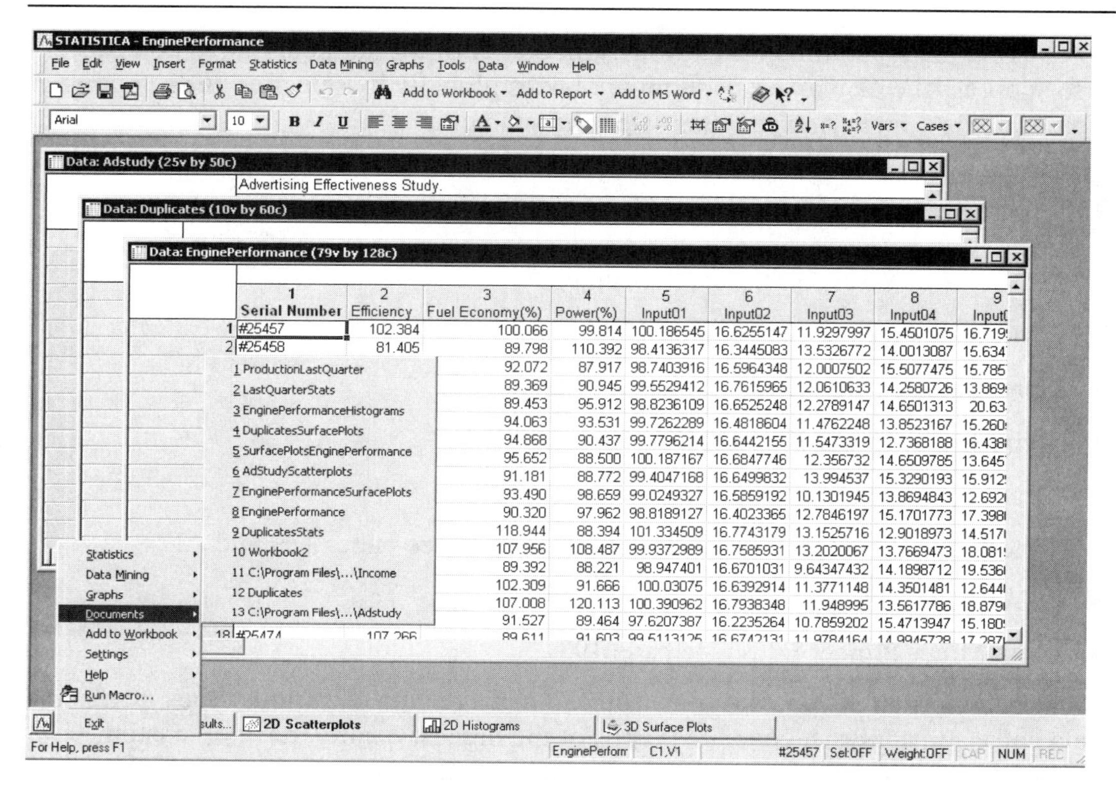

On the ***General*** tab of the ***Options*** dialog (accessible via the ***Tools - Options*** menu), you can specify how many recently used documents to display (the default is ***16***). For more detailed information about each document type, see the overviews for workbooks, spreadsheets, reports, graphs, and macros on page 177; for further information, see the *Electronic Manual*.

Toolbars related to types of active document windows. Each of the main types of *STATISTICA* document windows (see above) manages data in a different way and, thus, offers different customization and management options. These differences are reflected in the toolbars that accompany each type of window. Menu commands and toolbar buttons for each of the main types of documents are discussed in detail in the *Electronic Manual*.

Note that workbooks do not have a specialized toolbar (although the ***Standard*** toolbar is always available) because the toolbars that are available in workbooks depend on the type of document that is currently displayed in the workbook. Therefore, when you are editing a spreadsheet, graph, report, macro, or foreign document (e.g., an Excel spreadsheet) within a workbook, the toolbars and menus relevant for that document type are available.

When you select an "empty node" in the workbook tree pane, by default, the ***Statistics*** toolbar is displayed (in place of the specific document type toolbar) in order to maintain the same size and proportions of the application workspace.

User-defined toolbars. In addition to the variety of toolbars provided in *STATISTICA,* you can also create user-defined toolbars. These toolbars can include any command available in *STATISTICA,* as well as special controls (i.e., font name, font size, graph styles, etc.). The toolbars can be given any name and can be designated to open depending on the active document type. Also, you can customize all toolbars (including existing toolbars) by adding commands and special controls.

To create a toolbar (or edit an existing one) use the ***Toolbars*** tab of the ***Customize*** dialog accessible from the ***Tools - Customize*** menu. Customizing a toolbar is as easy as dragging commands from the dialog to the toolbar, as shown in the illustration below.

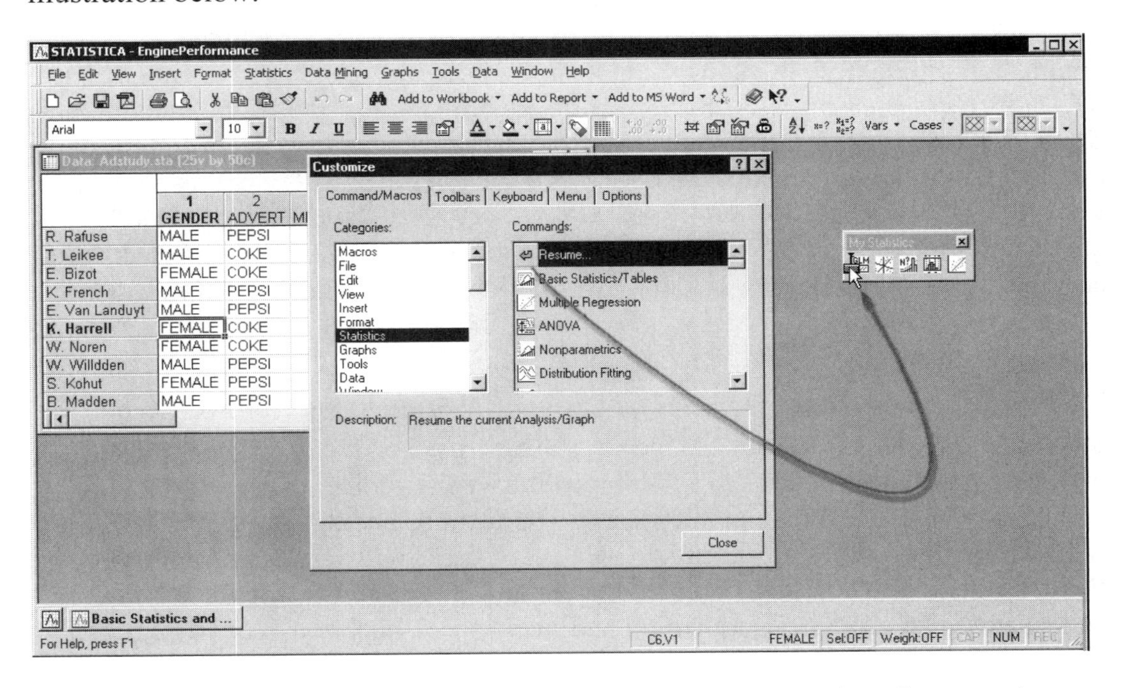

Shapes and locations of toolbars can be easily adjusted (e.g., all toolbars can be docked or free floating). All of these options make it possible for you to create

unique toolbars that provide you with a very specialized user interface. The *Electronic Manual* includes simple to follow, step-by-step instructions on how to make customizations. Specifically, see *Create New Toolbar* in the *Electronic Manual* for more details.

User-defined menus. Customizing the menus is equally easy and can be performed using the ***Menu*** tab of the ***Customize*** dialog (see the *Electronic Manual* for details).

STATISTICA VISUAL BASIC AND CONTROLLING *STATISTICA* FROM OTHER APPLICATIONS

The industry standard *STATISTICA* Visual Basic language (integrated into *STATISTICA*) provides an alternative user interface to the entire functionality of *STATISTICA*, and it offers incomparably more than just a "supplementary application programming language" that can be used to write custom extensions. *STATISTICA* Visual Basic takes full advantage of the object model architecture of *STATISTICA* and can be used to access programmatically every aspect and virtually every detail of the functionality of *STATISTICA*. Even the most complex analyses and graphs can be recorded into Visual Basic macros and later be run repeatedly or edited and used as building blocks of other applications. *STATISTICA* Visual Basic adds an arsenal of more than 13,000 new functions to the standard comprehensive syntax of Visual Basic, thus comprising one of the largest and richest development environments available. For more information on *STATISTICA* Visual Basic, see Chapter 9 (page 227).

Controlling *STATISTICA* from other applications. One of the features that makes the *STATISTICA* Visual Basic environment so powerful is the ability to integrate and manipulate various applications and their environments within a single macro. For example, you can record or write a *STATISTICA* Visual Basic program that computes predictions via the *STATISTICA* ***Time Series*** module and execute that program from within an Excel spreadsheet or a Word document. The exchange of information between different applications is accomplished by exposing those applications to the Visual Basic programs as Objects. So, for example, you can run statistical analyses in the *STATISTICA* ***Basic Statistics***

module from a Visual Basic program in Excel by declaring inside the program an object of type ***Statistica.Application***.

Once an object has been created, the Visual Basic program then has access to the properties and methods contained in that object. Properties can be mostly thought of as functions, methods can be mostly thought of as subroutines that perform certain operations or computations inside the respective application object. You can call *STATISTICA* procedures directly from many other applications and programming languages (e.g., C++, Java, and others).

WEB BROWSER-BASED USER INTERFACE: *WEBSTATISTICA*

In addition to the two basic types of user interfaces described in the previous sections, the entire *STATISTICA* family of products also optionally offers an Internet Browser-based user interface, where all interactions with the application involving querying databases, data management operations, data analysis, or data mining, as well as generating reports and collaborative work, can performed without having any *STATISTICA* application installed on the local computer, from virtually any computer in the world as long as it is connected to the Internet. This alternative user interface requires that a Client-Server version of the respective *STATISTICA* application be installed.

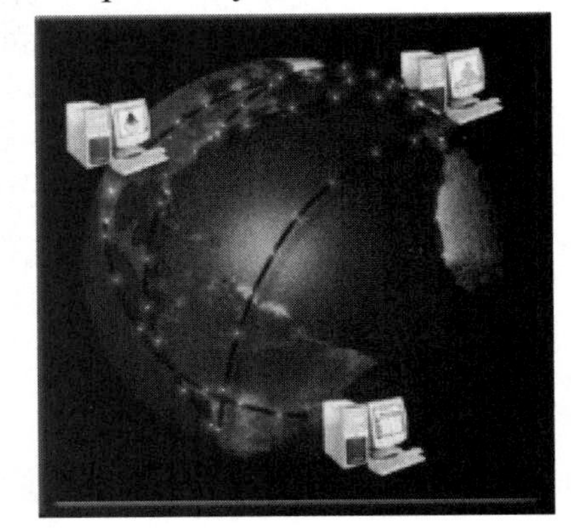

Overview (plain language). *WebSTATISTICA Server* adds full Internet enablement to *STATISTICA,* including the ability to interactively run *STATISTICA* from a Web browser. It enables you to easily and quickly access data and powerful analytical tools from virtually any computer in the world as long as it is connected to the Web. The product is provided with a selection of Internet browser-based user interfaces (in the form of extremely simple to navigate and easy-to-use dialogs), which enable you to specify analyses and review results using a standard Internet browser. Also, tools are provided to customize these dialogs and easily set up new, customized user interfaces or to add new functions. For example, a simple dialog with only three buttons can appear in the browser, where pressing each of the buttons will run a series of analyses and generate a detailed report. *WebSTATISTICA Server* applications add a new dimension and an endless array of possibilities to the entire line of *STATISTICA* Data Analysis, Data Mining, and Quality Control/Six Sigma software.

Overview (technical language). *WebSTATISTICA Server* is a highly scalable, enterprise-level, Web-based data analysis and database gateway application system, built on distributed processing technology and fully supporting multi-tier Client-Server architecture configurations. *WebSTATISTICA Server* exposes the analytic, query, reporting, and graphics functionality of *STATISTICA* through easy to use, interactive, standard Web interfaces. It is offered as a complete, ready to install application with an interactive, Internet browser-based ("point-and-click") user interface, allowing users in remote locations to interactively create data sets, run analyses, and review output. However, because of its open architecture, *WebSTATISTICA Server* also includes development kit tools (based entirely on industry standard syntax conventions such as VB Script, C++, HTML, XML), allowing IT departments to customize all main components of the system, or to expand it by building on its foundations, for example, by adding new components and/or corporation-specific analytic or database facilities. The system is compatible with all major Web server software platforms (e.g., UNIX Apache, Microsoft IIS), works in both Microsoft .NET and Sun/Java environments, and does not require any changes to the existing firewall and Internet/Intranet security systems.

For more information, please refer to *Appendix C – WebSTATISTICA*, page 269.

MICROSOFT OFFICE INTEGRATION

If Microsoft Office is installed on the same machine as *STATISTICA*, then Excel spreadsheets can be opened directly within *STATISTICA* and used as a data source for analyses, and Word documents can be used as a destination for reports (see example on page 55; see also page 158).

Excel as a data source. *STATISTICA* can open Excel documents in the *STATISTICA* workspace through the standard ***Open*** dialog. When an Excel workbook is selected, a dialog will be displayed that enables you to import the file

into a standard *STATISTICA* Spreadsheet or to keep the document in Excel form, i.e., as an Excel window within *STATISTICA*.

Once the Excel document is opened, you have access to all the menus and toolbars that Excel supports. Thus, you can edit and update formulas, change the formatting, copy/paste, drag/drop – everything that you would normally do if you were within the Excel application.

The main strength in Excel integration is that the Excel documents can be used as a data source for analyses. Simply have the Excel document window selected when starting an analysis, and the analysis will source from the Excel document. When initially running the analysis, *STATISTICA* will display a dialog in which you can specify what range of the Excel document should be used as the data source and if a particular row or column is to be used as variable names or case names. These settings are assigned to the Excel document so you will only need to specify them once.

Not only can *STATISTICA* use the Excel file as a data source, but auto updating can be specified as well. If you create an auto-updating graph from the ***Graphs*** menu, then change data in the Excel file by entering new data or re-evaluating formulas, the graph will also be updated.

Word as a report destination. You can also open and edit Word documents within the *STATISTICA* workspace. Word documents can be opened using the standard ***Open*** dialog, and when performing statistical analyses or creating graphs, output can being directed to a Word document. Any output that is capable of being directed to a *STATISTICA* Report is capable of being directed to a Word document.

As with Excel windows, Word windows contain all the toolbars and menus that are supported within the Word application. You can perform any formatting and editing that Word supports within its application.

When sending spreadsheet analytical results to Word, *STATISTICA* will take advantage of Word's table editing facility and convert the spreadsheet into a table. For multi-page spreadsheets, you can control where to break the rows and columns. These spreadsheets will be broken by columns such as will be allowed without exceeding the page width. All rows for a given set of columns will be rendered before the next set of spreadsheet columns is rendered in the Word document. This solution enables the presentation of spreadsheets in Word that are natively editable

in Word, display the entire contents of the spreadsheet, and print and paginate correctly.

FIVE CHANNELS FOR OUTPUT FROM ANALYSES

CHAPTER

FIVE CHANNELS FOR OUTPUT FROM ANALYSES

OVERVIEW

When you perform an analysis, *STATISTICA* generates output in the form of multimedia tables (spreadsheets) and graphs. There are five basic channels to which you can direct all output:

1. *STATISTICA* Workbooks (page 152)
2. Stand-alone Windows (page 154)
3. Reports (page 155)
4. Microsoft Word (page 158)
5. The Web (page 160)

The first four output channels listed above are controlled by the options on the ***Output Manager*** (accessible by selecting ***Output Manager*** from the ***File*** menu, see page 23 for further details on both the global ***Output Manager*** on the ***Options*** dialog and the ***Analysis/Graph Output Manager*** dialog). There are a number of ways to output to the Web, depending on the version of *STATISTICA* you have.

These means for output can be used in many combinations (e.g., a workbook and report simultaneously) and can be customized in a variety of ways. Also, all output objects (spreadsheets and graphs) placed in each of the output channels can contain other embedded and linked objects and documents, so *STATISTICA* output can be hierarchically organized in a variety of ways. Each of the *STATISTICA* output channels has its unique advantages, as described in the following sections. More comprehensive overviews of each of the document types associated with the respective channels of output are included in the next chapter (page 175).

The auto save and recovery features. All *STATISTICA* documents (i.e., input spreadsheets, workbooks, reports, and macros) that are likely to accumulate the results of your work (e.g., data entry, editing, or output collection) over an extended period of time support the ***Auto Save*** feature, which is configurable on the ***General*** tab of the ***Options*** dialog (accessible by selecting ***Options*** from the ***Tools*** menu). This facility will automatically save the contents of your work periodically (e.g., every 10 minutes) and, thus, give you the option to retrieve data that otherwise could have been lost in case of a power outage or a system failure.

1. *STATISTICA* WORKBOOKS

Workbooks are the default way of managing output (for more information, see page 177). Each output document (e.g., a *STATISTICA* Spreadsheet or Graph, as well as a Word or Excel document) is stored as a tab in the workbook.

Documents can be organized into hierarchies of folders or document nodes (by default, one is created for each new analysis) using a tree view, in which individual documents, folders, or entire branches of the tree can be flexibly managed.

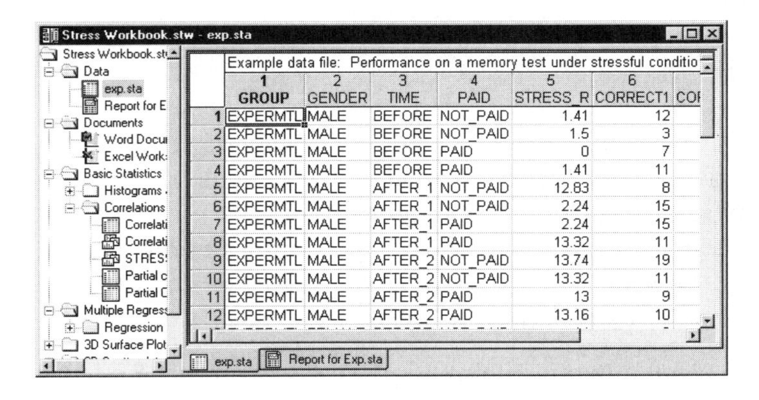

For example, selections of documents can be extracted (e.g., drag-copied or drag-moved) to a report window or to the application workspace (i.e., the *STATISTICA* application "background" where they will be displayed in stand-alone windows). Entire branches can be placed into other workbooks in a variety of ways in order to build specific folder organization, etc.

Technically speaking, workbooks are ActiveX document containers (see page 244 for information on ActiveX technology, see also the *Electronic Manual*). Workbooks are compatible with a variety of foreign file formats (e.g., Office documents) that can be easily inserted into workbooks and in-place edited.

User notes and comments in workbooks. Workbooks offer powerful options to efficiently manage even extremely large amounts of output, and they may be the best output handling solution for both novices and advanced users. It might appear that one possible drawback is that user comments (e.g., notes) and supplementary information cannot be as transparently inserted into the "stream" of the workbook output as they can in traditional, word processor style reports, such as *STATISTICA* Reports (see the next section). However, note that:

- All *STATISTICA* documents can easily be annotated, both a) directly, by typing text into graphs, tables, and reports, and b) indirectly, by entering notes into the ***Comments*** box of the ***Document Properties*** dialog (accessed from the ***File - Properties*** menu), and
- Formatted documents with notes and comments (in the form of text files, *STATISTICA* Report documents, WordPad or word processor documents, etc.) can easily be inserted anywhere in the hierarchical organization of output in workbooks. Moreover, such summary notes or comment documents can be made nodes for groups of subordinate objects to which the note is related to further enhance their organization.

Saving workbooks as Web pages. Workbooks can be saved as **.html* (Web) files by selecting ***Save As*** from the *STATISTICA* ***File*** menu, and in the ***Save As*** dialog, choosing ***Web Page (*.htm; *.html)*** from the ***Save as type*** drop-down list. Saving as a Web page will create an **.html* file in the specified directory that can be opened with standard internet browsers such as Microsoft Internet Explorer. When saving the workbook as a Web page, *STATISTICA* also creates a subdirectory that contains all the images referenced by the Web page.

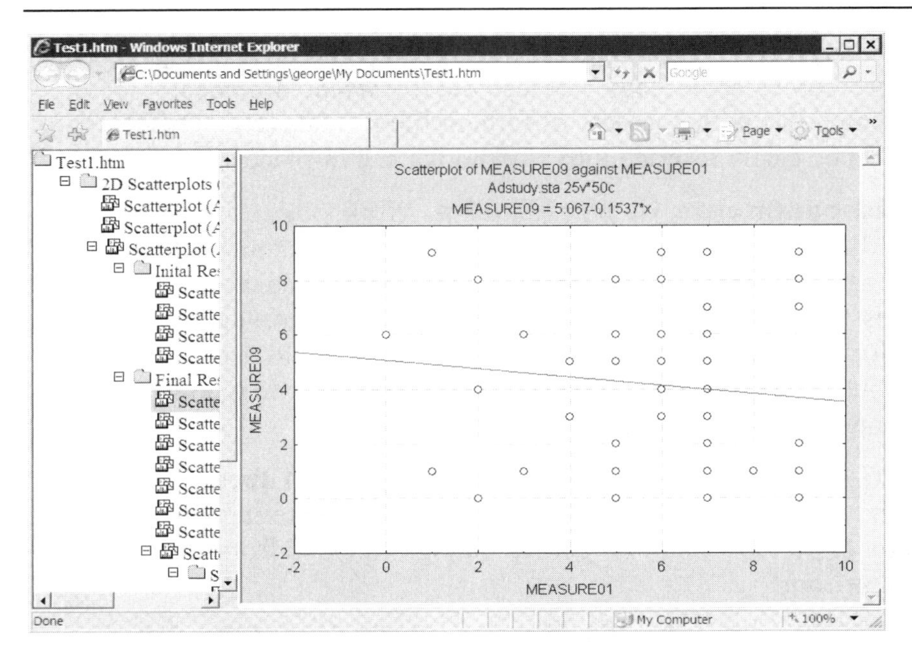

The Web page output contains an *.html*-based tree control that enables you to navigate and display the various workbook images, similar to the actual workbook.

2. STAND-ALONE WINDOWS

STATISTICA output documents can also be directed to a queue of stand-alone windows; the ***Queue Length*** can be controlled on the ***Output Manager*** tab of the ***Options*** dialog (accessible by selecting ***Output Manager*** from the ***File*** menu).

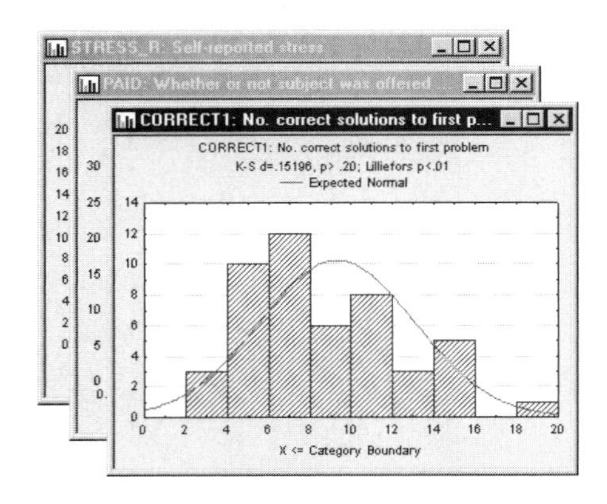

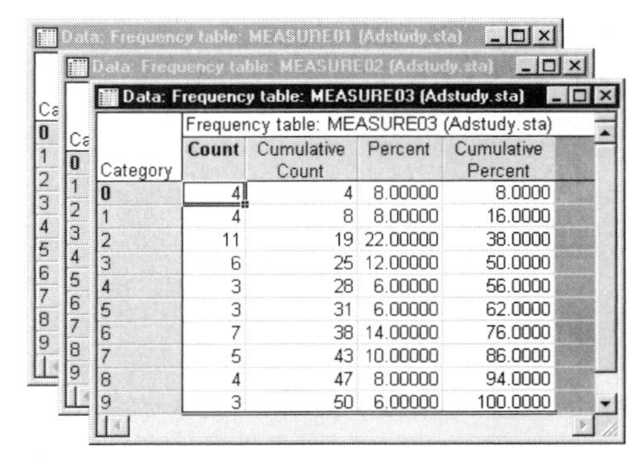

Frequency table: MEASURE03 (Adstudy.sta)

Category	Count	Cumulative Count	Percent	Cumulative Percent
0	4	4	8.00000	8.0000
1	4	8	8.00000	16.0000
2	11	19	22.00000	38.0000
3	6	25	12.00000	50.0000
4	3	28	6.00000	56.0000
5	3	31	6.00000	62.0000
6	7	38	14.00000	76.0000
7	5	43	10.00000	86.0000
8	4	47	8.00000	94.0000
9	3	50	6.00000	100.0000

The clear disadvantage of this output mode is its total lack of organization and its natural tendency to clutter the application workspace (note that some procedures can generate hundreds of tables or graphs with a click of the button).

One of the advantages of this way of handling output is that you can easily custom arrange these objects within the *STATISTICA* application workspace (e.g., to create multiple, easy to identify "reference documents" to be compared to the new output). However, note that in order to achieve that effect, you do not need to configure the output ahead of time and generate a large number of (mostly unwanted) separate windows that can clutter the workspace. Instead, individual, specific output objects directed to and stored in the other two channels (workbooks and reports) can easily be dragged out from their respective tree views onto the application workspace as needed.

3. REPORTS

When performing an analysis, the ultimate goal is to create meaningful output in order to gain an understanding of the data. The manner in which the output is produced is important as well. *STATISTICA* offers a variety of methods to produce reports that accommodate the diverse needs of users.

STATISTICA Reports

STATISTICA Reports (for more information, see page 185) offer a more traditional way of handling output where each object (e.g., a *STATISTICA* Spreadsheet or Graph, or an Excel spreadsheet) is displayed sequentially in a word processor style document.

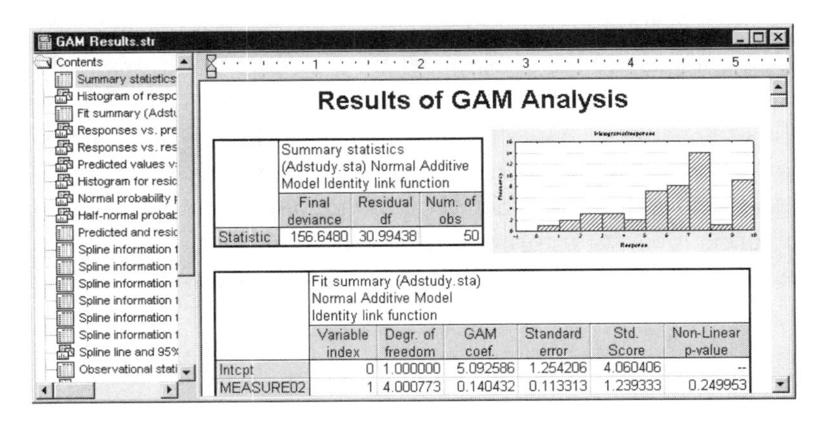

	Summary statistics (Adstudy.sta) Normal Additive Model Identity link function		
	Final deviance	Residual df	Num. of obs
Statistic	156.6480	30.99438	50

	Fit summary (Adstudy.sta) Normal Additive Model Identity link function					
	Variable index	Degr. of freedom	GAM coef.	Standard error	Std. Score	Non-Linear p-value
Intcpt	0	1.000000	5.092586	1.254206	4.060406	--
MEASURE02	1	4.000773	0.140432	0.113313	1.239333	0.249953

However, the technology behind this simple editor offers you very rich functionality. For example, like the workbook (see the previous section), the *STATISTICA* Report is also an ActiveX container (for information on ActiveX technology, see page 244 or the *Electronic Manual*) where each of its objects (not only *STATISTICA* Spreadsheets and Graphs, but also any other ActiveX-compatible documents, e.g., Excel spreadsheets) remains active, customizable, and in-place editable.

The obvious advantages of this way of handling output (more traditional than the workbook) are the ability to insert notes and comments "in between" the objects as well as its support for the more traditional way of quick scrolling through and reviewing the output to which some users may be accustomed (the editor supports variable speed scrolling and other features of the IntelliMouse). Also, only the report output includes and preserves the record of the supplementary information, which contains a detailed log of the options specified for the analyses (e.g., selected variables and their labels, long names, etc.) depending on the level of supplementary information specified on the ***Output Manager*** tab of the ***Options*** dialog (accessible by selecting ***Output Manager*** from the ***File*** menu), see page 24.

The obvious drawback, however, of these traditional reports is the inherent flat structure imposed by their word processor style format, although that is what some users or certain applications may favor.

Reports from Workbooks

When you have a *STATISTICA* Workbook containing analyses output, you may decide you want to transfer it to a report.

The advantages of this way of handling output are the ability to insert notes and comments in between the objects, as well as its support for the more traditional way of quick scrolling through and reviewing the output to which some users may be accustomed. Also, only the report output includes and preserves the record of the supplementary information, which contains a detailed log of the options specified for the analyses (e.g., selected variables and their labels, long names, etc.) depending on the level of supplementary information specified on the ***Output Manager*** tab of the ***Options*** dialog (accessible by selecting ***Options*** from the ***Tools*** menu), see page 24.

Open a *STATISTICA* Workbook. If you don't have a workbook saved, create a new one: first verify that the ***Workbook*** option button is selected in the ***Output Manager*** (accessible by selecting ***Output Manager*** from the ***File*** menu). Then, create a workbook by working through *Example 1: Correlations* (page 11) or *Example 2: ANOVA* (page 34).

In the open workbook, select all of the files, i.e., select the first file, press the SHIFT key on your keyboard, and select the last file. Then, click the ***Add to Report*** button on the toolbar. All the files in the workbook will be duplicated in a *STATISTICA* Report.

RTF (Rich Text Format) Reports

RTF (Rich Text Format) is a Microsoft standard method of encoding formatted text and graphics for easy transfer between applications. When reports are saved in Rich Text Format (**.rtf*), all file formatting is preserved so that it can be read and interpreted by other RTF-compatible applications (e.g., Word).

The *STATISTICA* Report format (*.str*) adheres to RTF conventions; however, saving reports in the default *STATISTICA* Report format ensures that the reports will be opened in *STATISTICA*, giving you complete access to the report tree.

In order to open a *STATISTICA* report in an RTF-compatible application, open the report and select ***Save As*** from the ***File*** menu to display the ***Save As*** dialog. From the ***Save as type*** drop-down list, select *Rich Text Files (*.rtf)*, enter a name in the ***File name*** field, and click the ***Save*** button. You can then open the file in any RTF-compatible application.

Acrobat (PDF) Reports

PDF is the acronym for Portable Document Format; it is the industry-standard format for storing textual and graphical data. PDF offers a graphically rich appearance and structure that makes it ideal for presentation purposes. Additionally, PDF documents can be viewed in both image and textual mode, i.e., you can either select data as a formatted image or as regular text.

PDF is platform independent, and most operating systems offer free PDF viewing applications (e.g., Adobe Acrobat on Windows and Ghostscript on Linux).

PDF has been approved as an acceptable document storage format for regulated environments according the FDA's 21 CFR Part 11.

To save a *STATISTICA* Report as a PDF file, open the report and then select ***Save As PDF*** from the ***File*** menu. The ***Output Options*** dialog will be displayed, where you can choose whether to output spreadsheets as ***Objects (as they are sized in the Report window)*** or ***Full-sized Spreadsheets (on separate pages)***. If you always want to output spreadsheets in the same manner, select the ***Use the current setting and do not display this dialog again*** check box. Click the ***OK*** button to close the ***Output Options*** dialog and display the ***Save report as PDF*** dialog. Use the ***Save in*** field to select the appropriate location in which to save the document, enter a name in the ***File name*** field, and click the ***Save*** button. *STATISTICA* Reports, Spreadsheets, and Graphs can all be saved in PDF format.

Note that, these are not simplified PDF files (representing compressed bitmaps of the respective document page images) but full-featured PDF files that support such operations as selective copying of text information.

4. MICROSOFT WORD

STATISTICA also allows routing output directly to Word via the Office Integration features. When Word is open within *STATISTICA,* Word toolbars and menus are also available through standard Active X Document interfaces technology. In *STATISTICA*, you can perform any formatting and editing that Word supports in its application.

When sending spreadsheet analytical results to Word, *STATISTICA* will take advantage of Word's table editing facility, and convert the spreadsheet to a table. For multi-page spreadsheets, you can control where to break the rows and columns. These spreadsheets will be broken by columns such as will be allowed without exceeding the page width. All rows for a given set of columns will be rendered before the next set of spreadsheet columns is rendered in the Word document. This solution enables the presentation of spreadsheets in Word that are natively editable in Word, displays the entire contents of the spreadsheet, and prints and paginates correctly.

As with standard *STATISTICA* Reports (see page 155), Word documents can store and preserve the record of supplementary information (e.g., selected variables, long names, etc.).

To send output to a Word document, use the options in the ***Output Manager*** (accessible by selecting ***Output Manager*** from the ***File*** menu, or by selecting ***Options*** from the ***Tools*** menu and selecting the ***Output Manager*** tab in the ***Options*** dialog). In the ***Microsoft Word Output*** drop-down list, select either ***Multiple Word documents (one for each analysis/graph)***, ***Common Word document (one shared for all analyses/graphs)***, or ***[Select File]*** to browse to a preexisting Word document.

Although Word documents do not provide the navigational tree of a *STATISTICA* Workbook or Report, the advantages in sending output to Word documents are many. By sending results to a Word document, you have all the word processing features of Word at your finger tips. For example, you can attach templates to create customized documents, add tables of content and indices, track changes, etc.

When inserting a large spreadsheet into a Word document, *STATISTICA* automatically detects how many variables can fit on each page and partitions the spreadsheet into several Word tables. If the spreadsheet uses case names, those names will be the first column in each table.

Additional benefits of sending results to a Word document include increased printing functionality (e.g., printing to files, manual duplex) and the ability to save results as Web pages

HTML Reports

You may want to post a *STATISTICA* Report or Workbook on the Internet for others to review. With *STATISTICA*, you can save reports and workbooks in HTML format. HTML is an acronym for HyperText Markup Language. HTML uses tags to identify elements of the document, such as text or graphics.

Open a *STATISTICA* Report or Workbook, and select ***Save As*** from the ***File*** menu to display the ***Save As*** dialog. From the ***Save as type*** drop-down list, select ***HTML files (*.html; *.htm)*** to save the file with an **.htm* extension.

Note that any graphs in the report or workbook are saved as **.png* files in the same folder as the HTM file using the following naming convention: *reportname_pict0001.png, reportname_pict0002.png*, etc. You can save graphs as JPG files, instead. To do this, select ***Options*** from the ***Tools*** menu to display the

Options dialog. Select either the ***Reports*** tab or the ***Workbooks*** tab, according to which document you intend to send to an *.htm* document, select the ***JPEG format*** option button in the ***Export HTML images as*** group box, and click ***OK***.

5. OUTPUT TO THE WEB

WebSTATISTICA Knowledge Portal

WebSTATISTICA provides another way to distribute reports through the Knowledge Portal. The Knowledge Portal enables you to publish *STATISTICA* documents (spreadsheets, graphs, reports, or workbooks) to the Internet. Users with limited Knowledge Portal permissions can then view those documents. You can control who can access these documents by setting permissions on the documents and directories using standard *WebSTATISTICA* repository tools.

To publish content in the Knowledge Portal, first create a directory in the *WebSTATISTICA* repository in the *Portal* folder: log in to *WebSTATISTICA* as a user with Administrator rights, and from the ***File*** menu, select ***My Directory Operations*** to display the *WebSTATISTICA* ***My Directory*** dialog; the content will look similar to the following illustration.

My Directory

Directories: Copy, Move, Rename, Delete, Security, Create

Datasets
DataMiner
Portal
My Directory

10Items.sta
2(4-0).sta
2K-P.sta
2level.sta
30PredictorsOfYield.sta
3x3.sta
4bar linkage.sta
Accident.sta
Accident2.sta
Accuracy.sta
Activities.sta
Adapters.sta
Adstudy.sta
Aggressn.sta

Files: Copy, Move, Rename, Delete, Security, Open, Download, Set Active, View
New Window

Show Empty Directories | Show Temporary Files | Show Jobs | Show All Users | Show System Files

Refresh

To create a folder in the *Portal* directory to contain your reports, select the *Portal* folder, and then click the ***Create*** button to display the ***Explorer User Prompt*** dialog. In the edit field, enter the new directory name of *Sample Portal Folder*, and click ***OK***. A dialog will be displayed confirming that *The directory /Portal/Sample Portal Folder was created*. Click the ***Show My Directory*** button, and you will be returned to the ***My Directory*** dialog. Select the ***Show Empty Directories*** check box, and then click the ***Refresh*** button. Expand the *Portal* directory by clicking the + next to that folder, and the new *Sample Portal Folder* will be displayed.

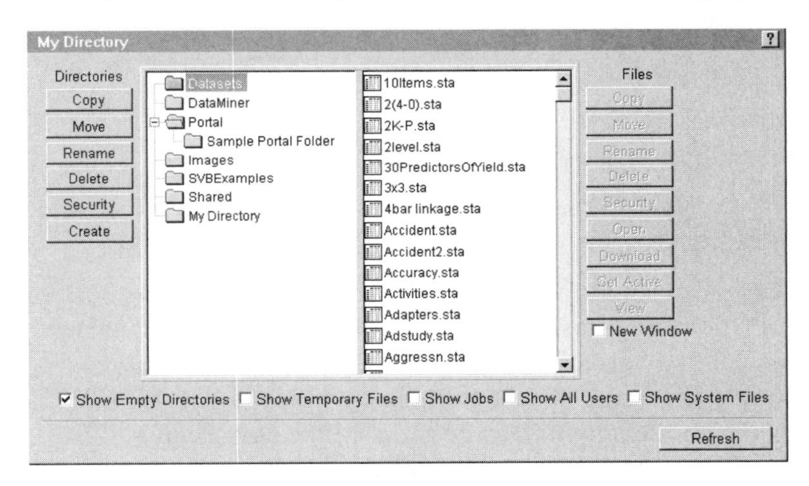

Note that you can control who can read and write to this folder by selecting the *Sample Portal Folder*, clicking the ***Security*** button, and using the *WebSTATISTICA* options to set the user and group permissions for the folder appropriately.

Publishing Content from *WebSTATISTICA*

Now that the folder has been created, you can add analysis results to it for *Portal* users to view using either *WebSTATISTICA* or *STATISTICA*.

In *WebSTATISTICA*, start with a typical analysis. From the *WebSTATISTICA* ***File*** menu, select ***Open Data Spreadsheet***. In the ***Select Data Source*** dialog, select the *Datasets* folder in the left pane, select the data file *Adstudy.sta* in the right pane, and click ***OK***.

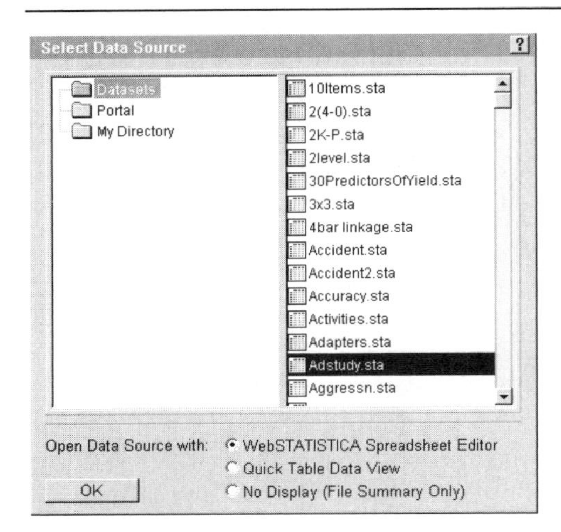

Close the resulting *WebSTATISTICA* Spreadsheet Editor window (we won't need it in this example), leaving just the browser window displaying the *Active Data source* summary information for *Adstudy.sta.*

From the *WebSTATISTICA* ***Statistics - Basic Statistics and Tables*** submenu, select ***Descriptive Statistics*** to display the variable selection dialog and the ***Descriptive Statistics*** parameters dialog. In the variable selection dialog, select *MEASURE01* and *MEASURE02* from the ***Continuous variables*** column. In the ***Descriptive Statistics*** parameters dialog, select ***All results*** in the ***Detail of computed results reporte***d field.

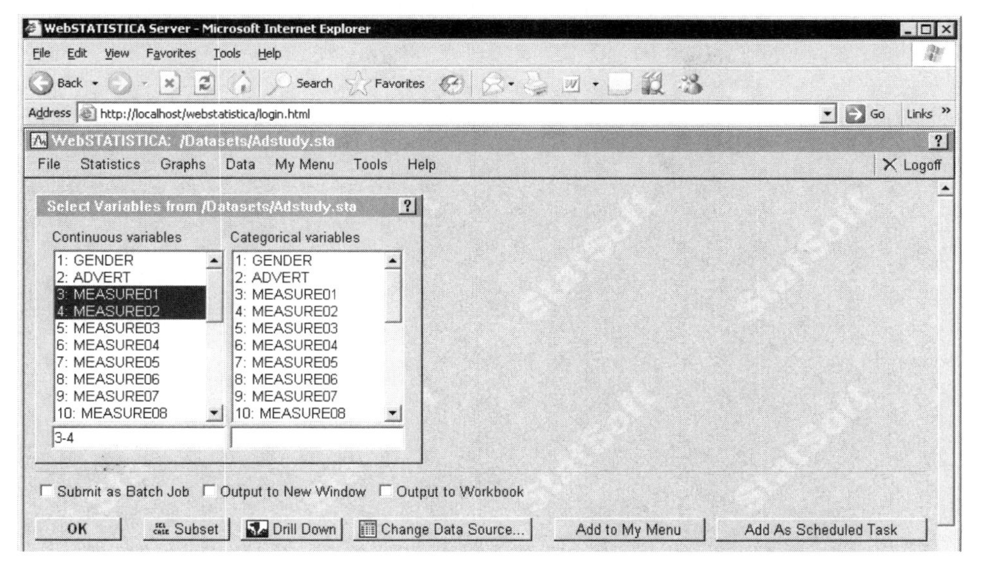

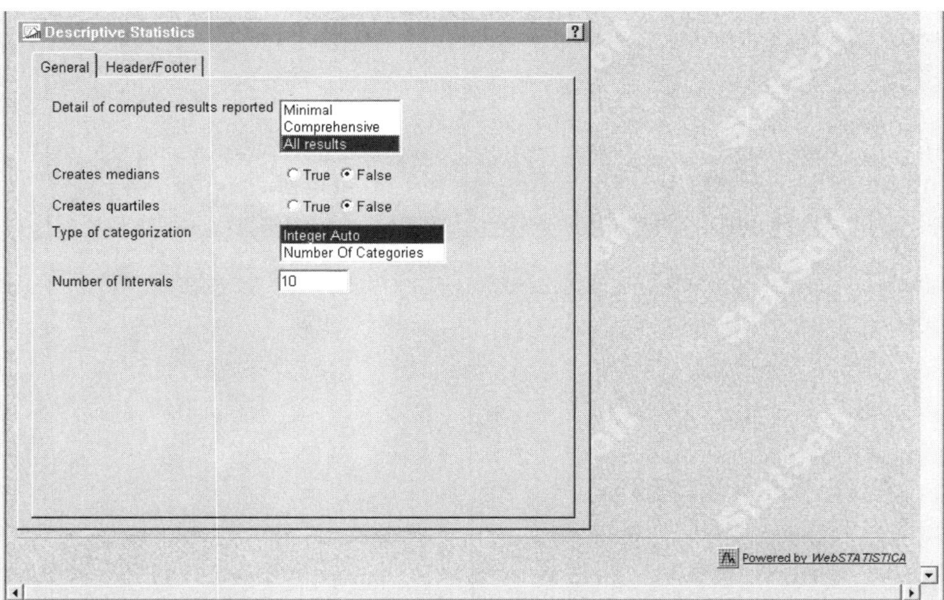

Click ***OK*** to display the results for this analysis, consisting of several spreadsheets and graphs.

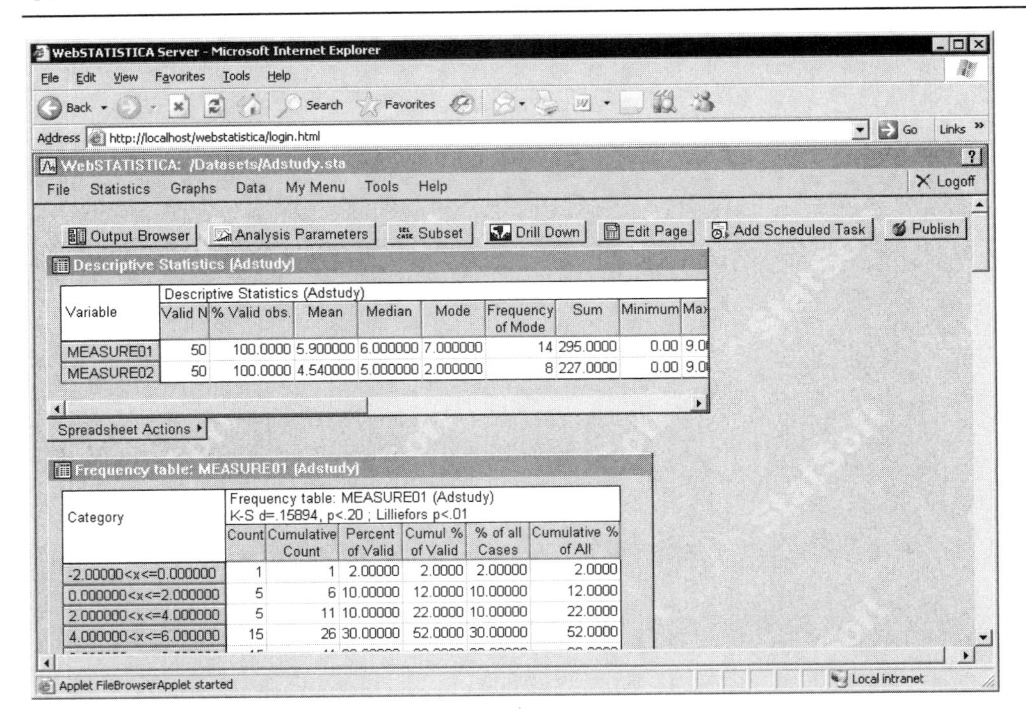

Before publishing this page to the Knowledge Portal, you may want to rearrange objects or annotate with text. Click the ***Edit Page*** button near the top of the page. This will open the ***Knowledge Portal Interactive Results Editor***.

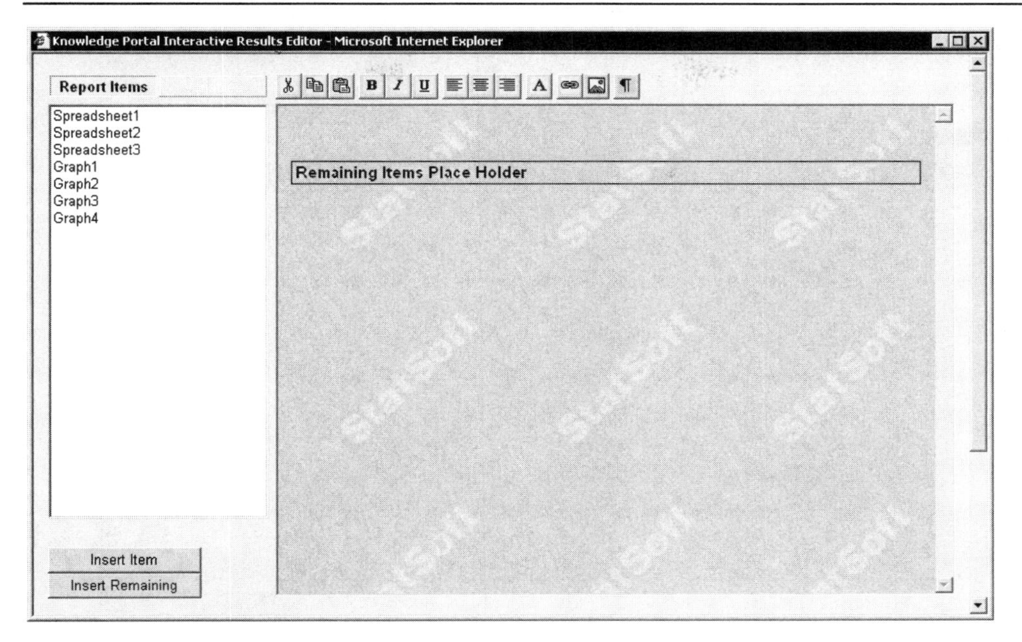

On the left side of the ***Editor*** is the list of objects found on the page; in this case, it is three spreadsheets and four graphs. The right side of the page is a representation of the content on the page. The initial object in the page is ***Remaining Items Place Holder***; this means that all items that have not specifically been selected will be included at this point in the page. Since this is the only item currently in the page, it means that the three spreadsheets and four graphs will be included in the output, just as it was on the original page.

For this example, let's include only Graph2 and Spreadsheet 3 in the results, and in that order. First, select the ***Remaining Items Place Holder*** object in the page view, and delete it. Now the page content is completely blank. Select ***Graph2*** from the list on the left, and click the ***Insert Item*** button to insert the graph on the page. To place the spreadsheet next, first click underneath the graph to move the insert marker under the newly added graph. Then select ***Spreadsheet3*** in the list on the left, and click the ***Insert Item*** button. The spreadsheet will be placed underneath the graph. Notice how the items that are currently included on the page are selected in the list control on the left.

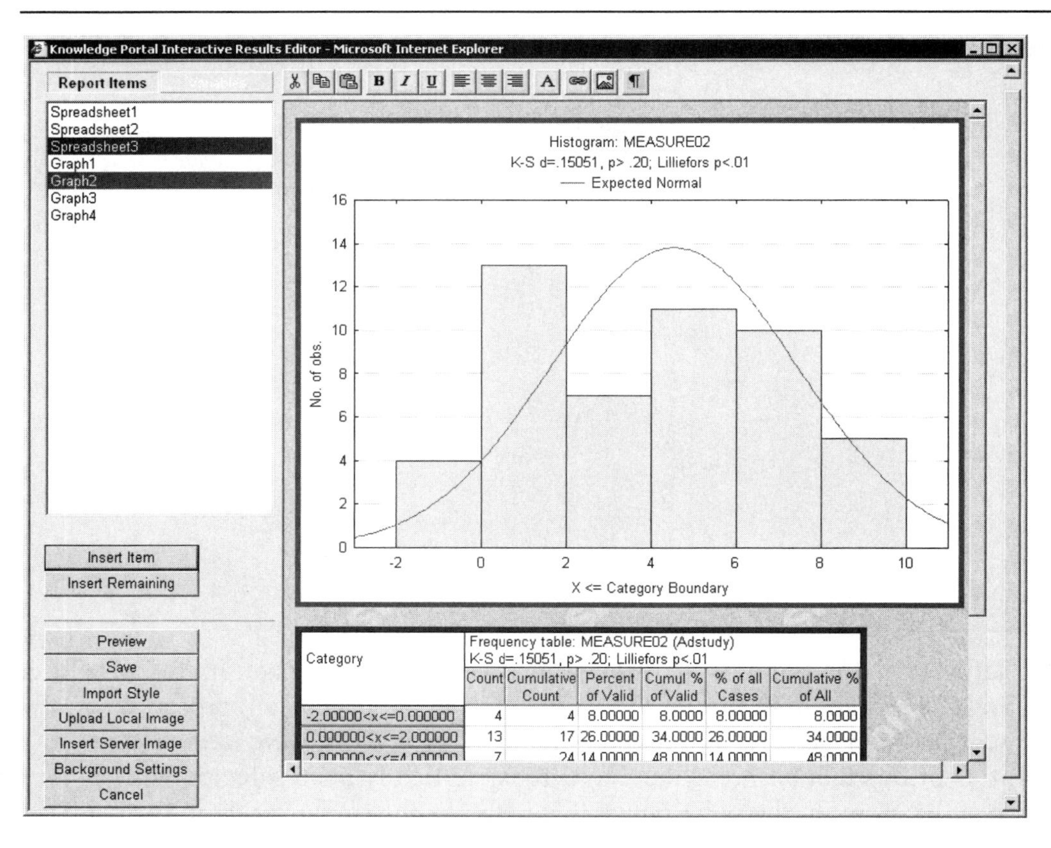

Let's annotate this with some text between the two objects. Place the cursor between the two objects in the page view, and type the text: *This is Frequency Table results of MEASURE02 from*. To change the font and point size, select all the text you just entered, and click the ***A*** toolbar button. Select a style of ***Bold*** and a size of ***24***, and click ***OK***.

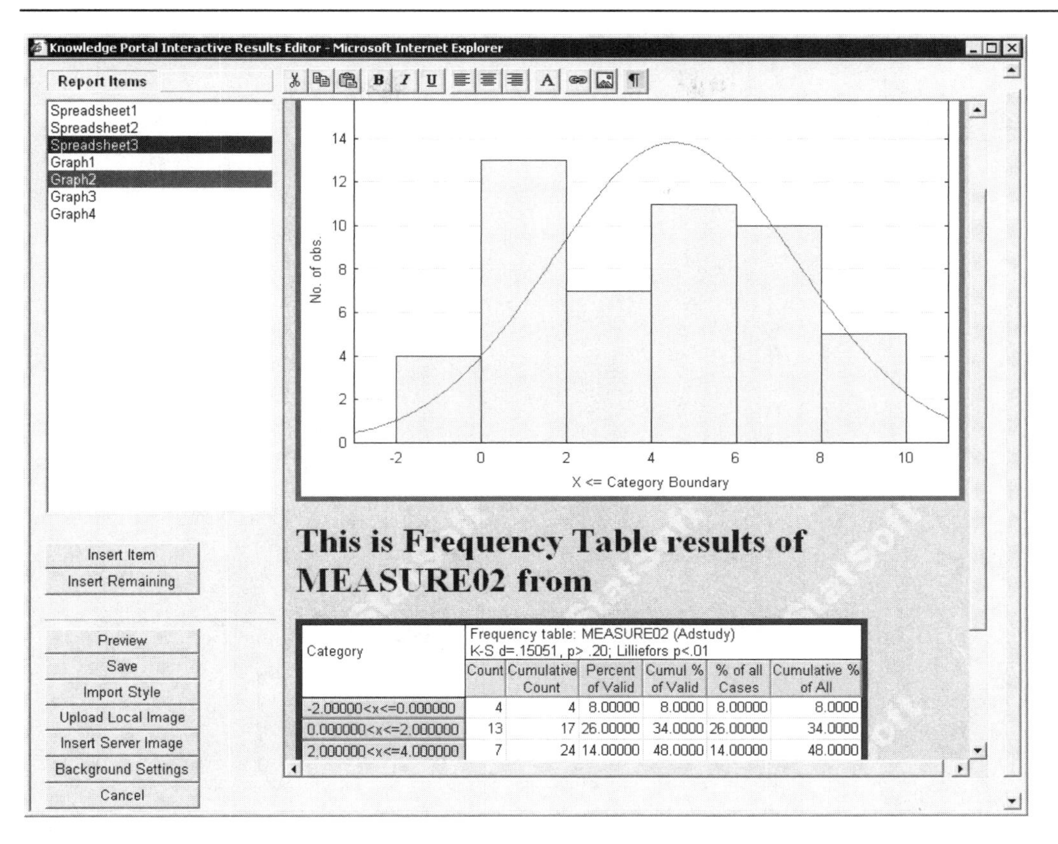

In addition to the objects that can be inserted, you can also insert predefined tags for this analysis. This includes all the parameters from the initial parameters dialog, as well as tags that represent the current date and time, current user, and name of the data source. To access these tags, click the ***Variables*** button in the upper-left of the ***Knowledge Portal Interactive Results Editor***, and the list control will display the tags that can be inserted. Place your cursor at the end of the text you just entered, click the *SWS_DataSource* tag, and click the ***Insert Variable*** button. This will insert the current name of the data source at the current cursor location.

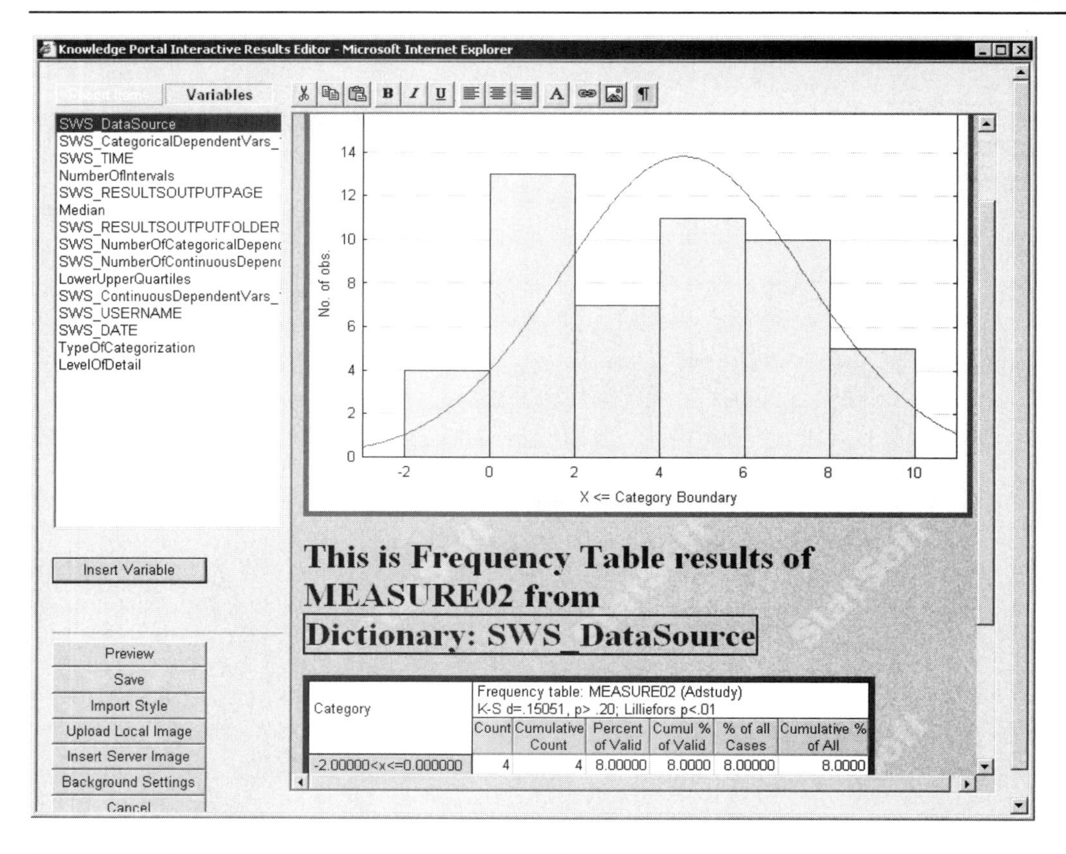

Note that the tag is a special marker and does not include the actual name of the data source in the ***Editor***, but the data source name will be shown when the page is displayed outside the ***Editor***. To display a preview of what the page will look like, click the ***Preview*** button. To save the modified page back to the original page, click the ***Save*** button. A ***Save Succeeded*** message will be displayed.

Click ***OK*** on this message, and then close the ***Knowledge Portal Interactive Results Editor***. You will see that your original page has been updated with the changes you made.

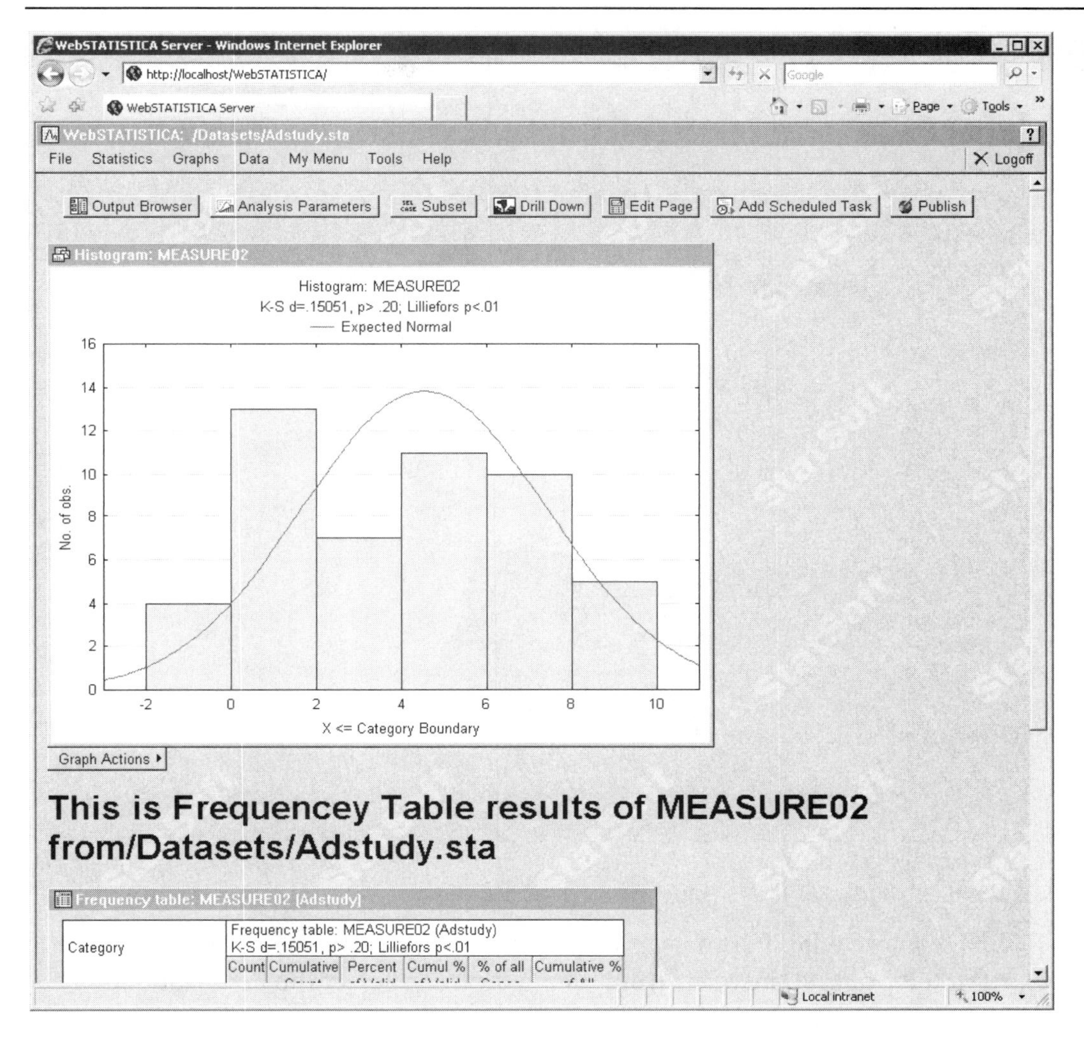

Now, to publish this page so that other users can see it from the Knowledge Portal, click the ***Publish*** button in the upper-right portion of the window. The ***Publish Destination*** dialog will be displayed. Here you can select the *Sample Portal Directory* that you created. You also can control who can have access to this particular page by selecting the ***I want to define who can access this output page*** check box; for this example, leave the check box cleared so all users who have access to the *Sample Portal Directory* will be able to see this document.

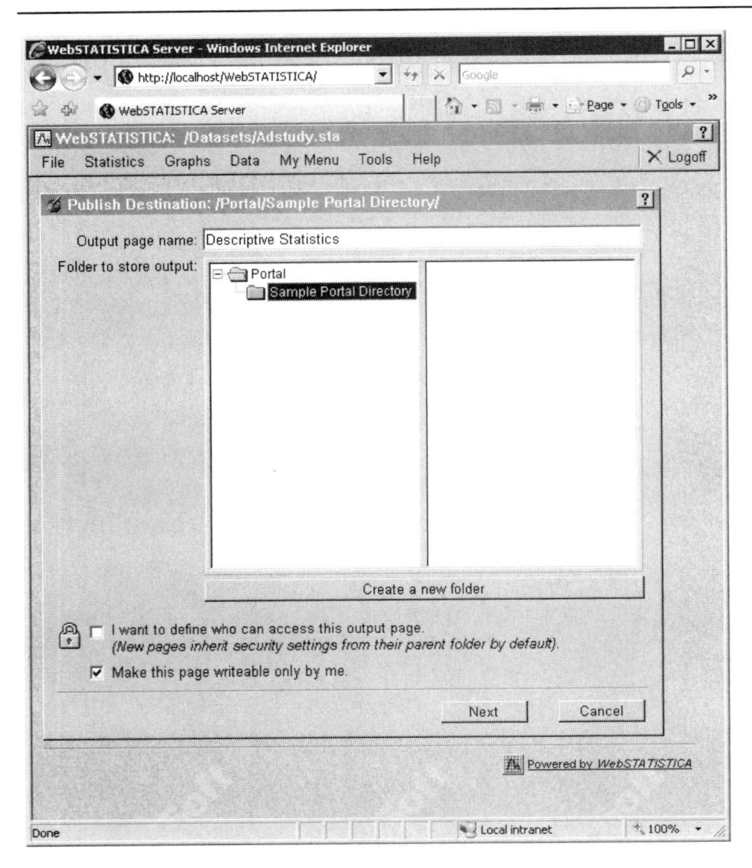

Click the ***Next*** button, and the page will be saved to the selected destination.

Now, when a Knowledge Portal user logs on, they will see the new *Sample Portal Directory* in their output browser, from which they can select the newly added *Descriptive Statistics* page.

Publishing Content from *STATISTICA* Desktop Applications

With the *WebSTATISTICA* integration feature of desktop *STATISTICA*, you can also publish *STATISTICA* documents (spreadsheets, graphs, reports, and workbooks) to the Knowledge Portal directly from within the *STATISTICA* application.

The first step is to enable *WebSTATISTICA* integration. From the *STATISTICA* ***Tools*** menu, select ***Options*** to display the ***Options*** dialog. Select the ***Server / Web*** tab, where you can specify the URL of the *WebSTATISTICA* Server and any optional custom configuration settings that may have been defined by your system administrator when installing *WebSTATISTICA*. In the following illustration, *WebSTATISTICA* has been installed on *serverx23*; the information in your dialog will be different depending on where *WebSTATISTICA* is installed on your network.

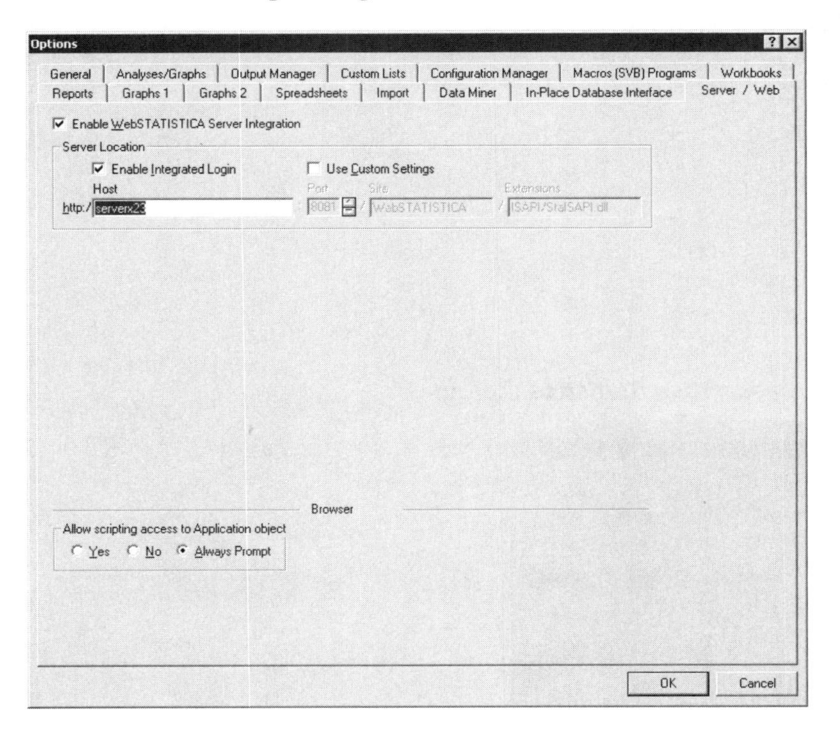

After you click the ***OK*** button in the ***Options*** dialog, note that there is a now a ***Server*** menu displayed in *STATISTICA* next to the ***File*** menu. The only command on the ***Server*** menu that is available initially is ***Log In***; select that command from the menu. If you have enabled integrated log in (and your Windows account is enabled on *WebSTATISTICA*), then you will be logged in automatically. Otherwise, you will be prompted for a *WebSTATISTICA* user name and password. Once you have logged in, the other commands are available on the ***Server*** menu.

Now, we will create an analysis and upload the results to the Knowledge Portal. Open the *Adstudy.sta* data file by selecting ***Open Examples*** from the ***File*** menu; in the ***Open a STATISTICA Data File*** dialog, double-click on the *Datasets* folder, and then double-click on the *Adstudy.sta* file to open that spreadsheet for use in *STATISTICA*. Now, from the ***Statistics*** menu, select ***Basic Statistics/Tables*** to display the ***Basic Statistics and Tables*** Startup Panel, and select ***Descriptive statistics***.

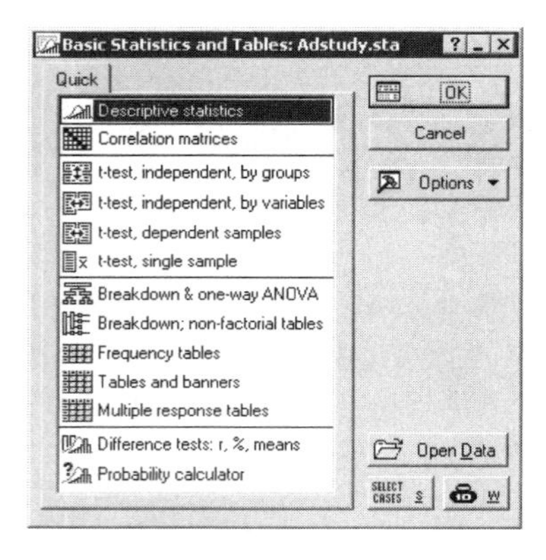

Click ***OK*** to display the ***Descriptive Statistics*** dialog.

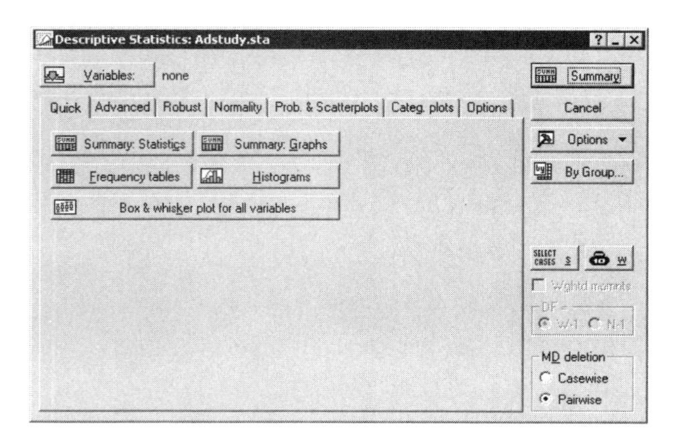

StatSoft

To ensure that all the output from this analysis will be sent to a workbook, click the ***Options*** button on the right side of the dialog, and from the drop-down list, select ***Output***. In the ***Analysis/Graph Output Manager***, verify that the ***Workbook*** option button is selected in the ***Place all results (Spreadsheets, Graphs) in*** group box. Then click ***OK*** to return to the ***Descriptive Statistics*** dialog.

Click the ***Variables*** button to display the variable selection dialog, select *MEASURE01* and *MEASURE02*, and click ***OK*** to return to the ***Descriptive Statistics*** dialog. On the ***Quick*** tab, click the ***Summary Statistics*** button to send those results to the workbook. The ***Descriptive Statistics*** dialog will be minimized so you can see the results; restore it by clicking the ***Descriptive Statistics*** button on the Analysis Bar on the lower-left of the screen. Now click the ***Histograms*** button to generate histograms for each selected variable. The analysis dialog is minimized again, and the workbook should look as follows.

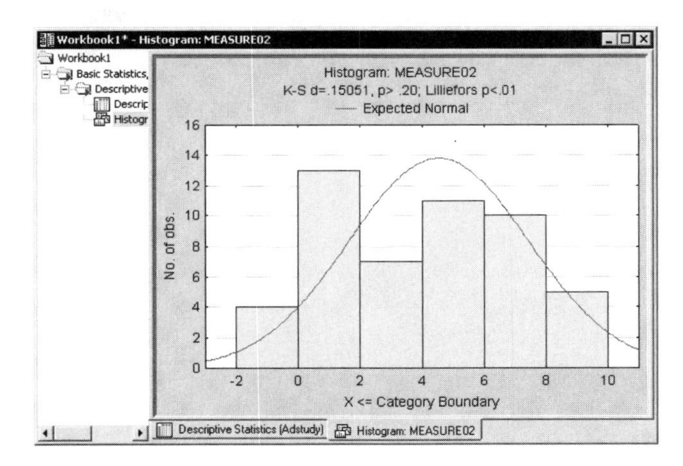

This is the document we want to publish to the Knowledge Portal. From the ***Server*** menu, select ***Save As***. The ***WebSTATISTICA Repository*** dialog will be displayed, containing a list of folders you can reference in *WebSTATISTICA*. Open the *Portal* folder, select the *Sample Portal Directory* folder, and click the ***OK*** button. This will upload the workbook to that Knowledge Portal directory.

You can review the document from within *STATISTICA* by opening a browser window inside of the *STATISTICA* workspace. From the ***Server*** menu, select ***Open In Browser***, and a new browser window will be opened, allowing you to log in to *WebSTATISTICA*. From the *WebSTATISTICA* ***File*** menu, choose ***My Directory Operations***; in ***My Directory***, you can navigate to the *Sample Portal Directory*, and see the *Workbook1.stw* file that was uploaded. Select this file and click the ***View*** button, and the workbook will be opened within the browser.

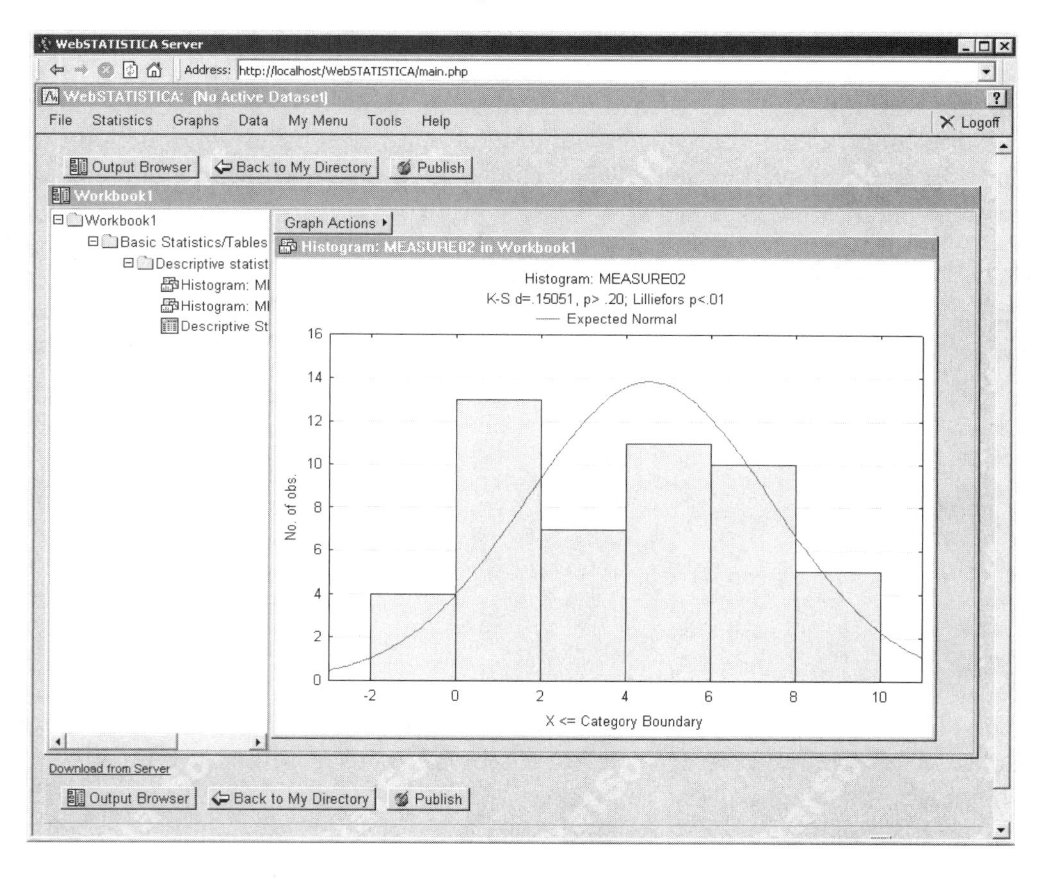

6

CHAPTER

STATISTICA DOCUMENTS

STATISTICA DOCUMENTS

WORKBOOKS

Workbooks (introduced briefly on page 152) are the default way of managing output. They store each output document (e.g., a *STATISTICA* Spreadsheet or Graph, as well as a Word or Excel document) as a tab.

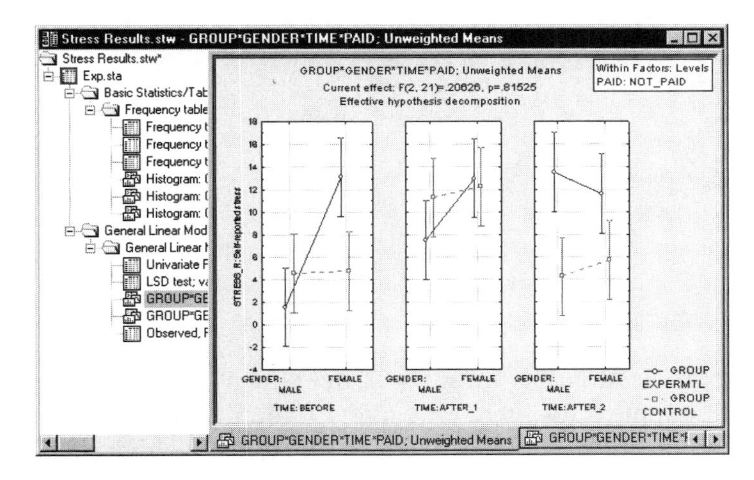

Technically speaking, *STATISTICA* Workbooks are optimized ActiveX (see page 244) containers that can efficiently handle large numbers of documents. The documents can be organized into hierarchies of folders or document nodes (by default, one is created for each new analysis) using a tree view, in which individual documents, folders, or entire branches of the tree can be flexibly managed.

For example, selections of documents can be extracted (e.g., drag-copied or drag-moved) to the report window or to the application workspace (i.e., the *STATISTICA* application "background" where they are displayed in stand-alone windows). Entire branches can be placed into other workbooks in a variety of ways in order to build a specific folder organization, etc.

Each workbook contains two panels: an Explorer-style navigation tree on the left and a document viewer on the right. The navigation tree (workbook tree) can be split into various nodes that are used to organize files in logical groupings (e.g., all analysis outputs or all macros created for a project). Tabs at the bottom of the document viewer (workbook viewer) are used to easily navigate the children of the currently selected node. You can easily move the tabs to the top, right, or left of the workbook viewer by right-clicking on one of the tabs and selecting a different location from the shortcut menu. One advantage of the side placement of tabs is that multiple rows (rather than one long row) are provided (as shown below). This makes it easy to select the desired tab.

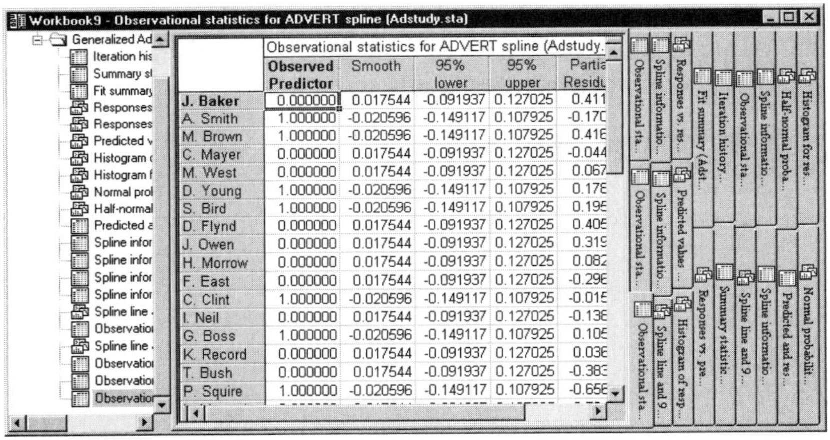

Workbook9 - Observational statistics for ADVERT spline (Adstudy.sta)

Observational statistics for ADVERT spline (Adstudy.

	Observed Predictor	Smooth	95% lower	95% upper	Partia Residu
J. Baker	0.000000	0.017544	-0.091937	0.127025	0.411
A. Smith	1.000000	-0.020596	-0.149117	0.107925	-0.170
M. Brown	1.000000	-0.020596	-0.149117	0.107925	0.416
C. Mayer	0.000000	0.017544	-0.091937	0.127025	-0.044
M. West	0.000000	0.017544	-0.091937	0.127025	0.067
D. Young	1.000000	-0.020596	-0.149117	0.107925	0.178
S. Bird	1.000000	-0.020596	-0.149117	0.107925	0.195
D. Flynd	0.000000	0.017544	-0.091937	0.127025	0.405
J. Owen	0.000000	0.017544	-0.091937	0.127025	0.319
H. Morrow	0.000000	0.017544	-0.091937	0.127025	0.082
F. East	0.000000	0.017544	-0.091937	0.127025	-0.298
C. Clint	1.000000	-0.020596	-0.149117	0.107925	-0.015
I. Neil	0.000000	0.017544	-0.091937	0.127025	-0.138
G. Boss	1.000000	-0.020596	-0.149117	0.107925	0.105
K. Record	0.000000	0.017544	-0.091937	0.127025	0.038
T. Bush	0.000000	0.017544	-0.091937	0.127025	-0.383
P. Squire	1.000000	-0.020596	-0.149117	0.107925	-0.658

Displaying tabs can also be suppressed to save the space. Unlike many Explorer-style navigation and organization applications that only allow folders to have children, the *STATISTICA* Workbook allows any item in the tree to have children. For example, you can add a spreadsheet to your workbook, and then add all the graphs produced using the data in the spreadsheet as children to the spreadsheet. A variety of drag-and-drop features and Clipboard procedures are available to aid you in organizing the workbook tree.

The workbook can hold all native *STATISTICA* documents including spreadsheets, graphs, reports, and macros. It can handle other types of ActiveX documents as well, including Excel spreadsheets, Word documents, and others. If you want to edit these documents, you can do so using the workbook viewer pane. To edit a Word document, double-click on the object in the workbook tree. The Word document opens in the viewer, and the workbook menu bar merges with the Word menu bar giving you access to all of the editing features you need. Workbooks can also be used to store all output from a particular analysis.

Navigating the Workbook Tree

The workbook tree displays the organization of files and folders in the workbook. The files and folders are displayed in an Explorer-style format. Items with plus signs next to them indicate folders or files that have children associated with them. To expand the tree for a particular folder or file, click the plus sign next to it. The workbook can support an unlimited number of levels, and both individual items from the tree view and entire branches can be flexibly (interactively) managed (e.g., right-click dragging to copy or move between workbooks or reports, etc., or via the shortcut menu, as shown below in the second image).

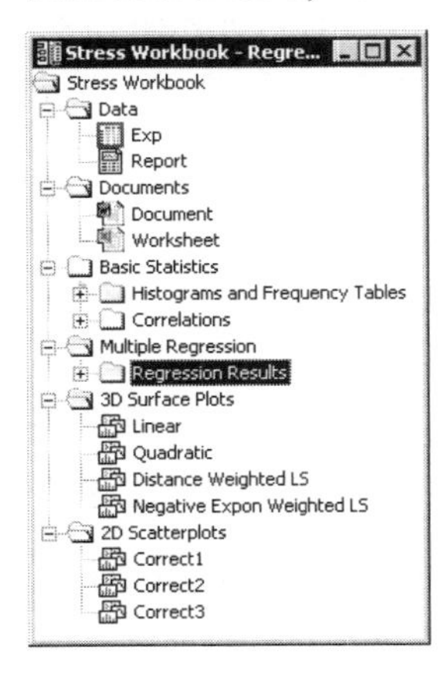

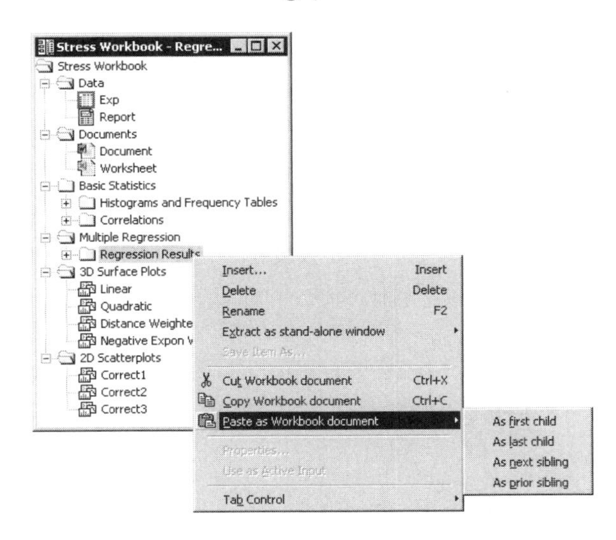

To select a workbook item for review or editing, simply locate the file in the workbook tree and click on its associated icon. The document will be displayed in the workbook viewer pane. Note that you can also navigate through the children of

the currently selected node using the navigation tabs available (by default) at the bottom of the workbook viewer. As mentioned previously, you can easily move these navigation tabs to the top, right, or left of the workbook viewer by right-clicking on one of the tabs and selecting a different location from the shortcut menu or selecting the appropriate command from the ***Workbook - Tab Control*** submenu. Note that tabs at the top and bottom of the viewer scroll sideways, while multiple rows of tabs are used when tabs are placed to the left or right of the viewer.

Items in the tree are identified by the icon next to them. The folder icon represents a folder that can contain a variety of documents and subfolders. The folder icon with a red arrow on it indicates that the script that produced the results in that folder has been attached to the folder. This enables *STATISTICA* to re-run or resume the analysis (for more details, see Chapter 9 – *STATISTICA Visual Basic*). The spreadsheet, report, macro, and graph icons represent *STATISTICA* Spreadsheet, Report, Macro, and Graph documents, respectively. The Data Miner icon represents a Data Miner workspace.

All non-*STATISTICA* documents are represented by their respective document icons. For example, Word documents are represented by the Word icon, and Excel spreadsheet files are represented by the Excel spreadsheet icon.

The workbook tree can be organized and modified using drag-and-drop features as well as Clipboard procedures. See *Workbook Drag-and-Drop Features* and *Workbook Clipboard Features* in the *Electronic Manual*. Commands for inserting, extracting, renaming, and removing items from the workbook tree are available from the workbook tree shortcut menu (accessed by right-clicking anywhere in the tree).

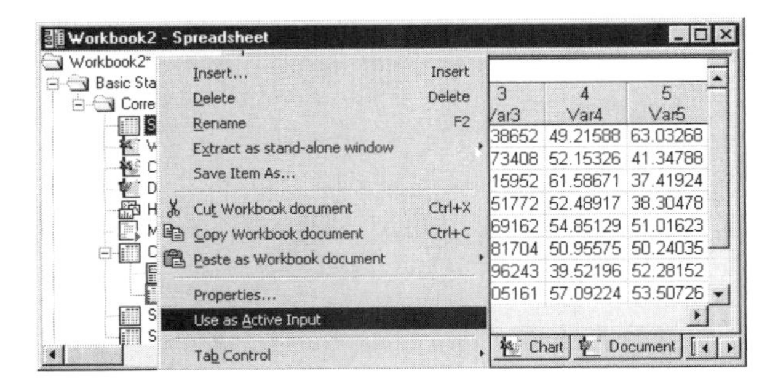

These commands are also accessible from the ***Workbook*** menu.

SPREADSHEETS (MULTIMEDIA TABLES)

STATISTICA Spreadsheets are based on StatSoft's proprietary multimedia table technology and are used to manage both input data and the numeric or text (and optionally any other type of) output. The basic form of the spreadsheet is a simple two-dimensional table that can handle a practically unlimited number of cases (rows) and variables (columns), and each cell can contain a virtually unlimited number of characters. Sound, video, graphs, animations, reports with embedded objects, or any ActiveX compatible documents can also be attached.

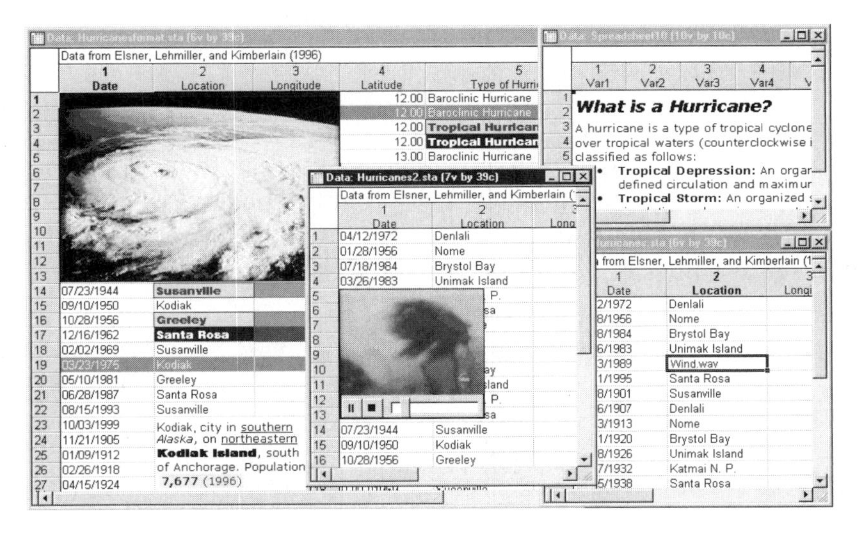

Because *STATISTICA* Spreadsheets can also contain macros and any user-defined user interface, these multimedia tables can be used as a framework for custom applications (e.g., with a list box of options or a series of buttons placed in the upper-left corner), self-running presentations, animations, simulations, etc.

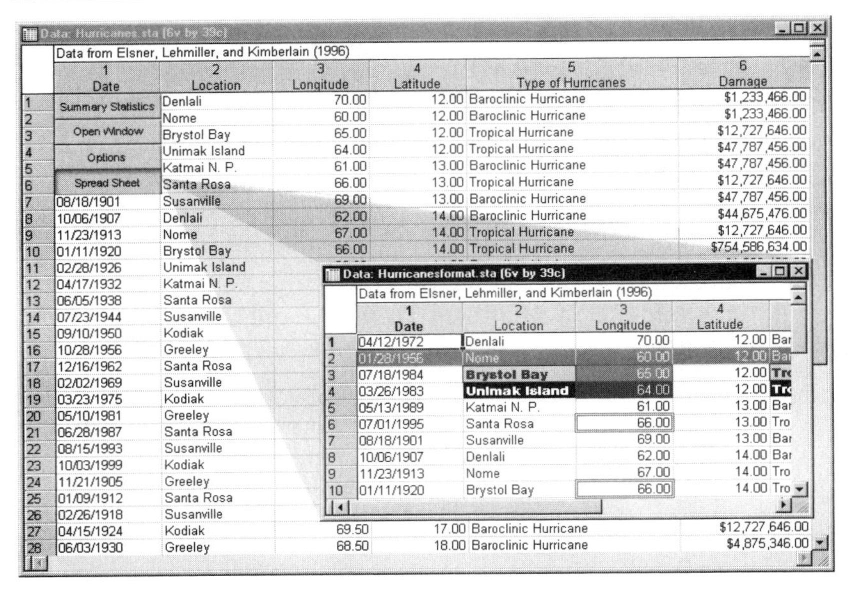

Data file layout in spreadsheets. *STATISTICA* data are organized into cases and variables. If you are unfamiliar with this notation, you can think of cases as the equivalent of records in a database management program (or rows of a spreadsheet), and variables as the equivalent of fields (or columns of a spreadsheet). Each case consists of a set of values of variables, and the first column in the file can (optionally) contain names of cases.

The spreadsheet window comprises several basic components.

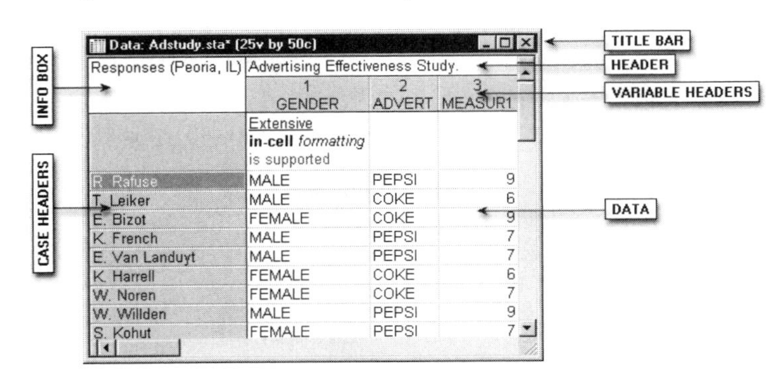

Title bar. The title bar displays the name of the spreadsheet followed by the spreadsheet extension (*.sta*). If the spreadsheet is an input spreadsheet, the title bar

also displays the number of variables by number of cases (e.g., *25v by 50c*). In the image shown above, the title bar contains the text *Data: Adstudy.sta (25v by 50c)*.

Info box. You can select the entire spreadsheet by clicking once in the lower-right corner (the mouse pointer will be the default arrow) of the info box, which is located in the upper-left corner of the spreadsheet window. To select the info box only (for formatting), click once in the upper-left corner of the info box (the mouse pointer will be an outlined plus sign ✥). Double-click in the info box to enter or edit the text in the info box (e.g., additional details about the spreadsheet). In the image shown above, the info box contains the text *Responses (Peoria, IL)*.

Header. The header is located immediately above the variable headers at the top of the window. Double-click the header to enter or edit text information. To select the header only (for formatting), click once in the upper-left corner (the mouse pointer will be an outlined plus sign ✥). Press CTRL+ENTER or ALT+ENTER to enter a new line (note that you need to extend the height of the field to see new lines that you are adding). In the image shown above, the header contains the text *Advertising Effectiveness Study*.

Case headers. These cells, located at the far left of the window, contain header information for each case. Double-click on any case header cell to enter or edit text information. To select the case header only (for formatting) click once on the left side of the case header (the mouse pointer will be an outlined plus sign ✥). To select the case row (for editing), click once on the middle or right side of the case header (the mouse pointer will be an outlined plus sign with an arrow ✥>). To select a block of case headers, (without selecting their respective rows), click on the left side of a case header and drag the mouse pointer to include all desired case headers. To autofit the case headers, double-click on the far-right side of any case header (the mouse pointer will be a cross with a double-headed arrow ↔). In the previous image, the case header cells contain the first initials and last names of the respondents in the study. Note that case headers are optional and you can choose not to display them (toggle off the ***Display Case Names*** command on the ***View*** menu); if they are not displayed, the case numbers are shown.

Variable headers. These cells, located at the top of each column, contain header information for each variable. To display details about an individual variable, double-click on the variable header cell. To select the variable header only (for formatting) click once on the upper portion of the variable header (the mouse pointer will be an outlined plus sign ✥). To select the variable column (for editing) click once on the lower portion of the variable header (the mouse pointer will be an

outlined plus sign with an arrow). To autofit the variable column, double-click on the far-right side of the variable header (the mouse pointer will be a cross with a double-headed arrow). In the previous image, the first two variable header cells contain the text *GENDER* and *ADVERT*. You have the option to change how the variable header cells display information so that they show the column number associated with the variable, the variable long name, and/or an abbreviation of the display types for the variables in the spreadsheet. Each of these options is available from the ***View - Variable Headers*** submenu.

Data (and in-cell formatting options). The remainder of the spreadsheet contains data that pertain to the cases and variables and any optional attached or linked objects (multimedia objects, macros, custom user interface). Text in cells can be of practically unlimited length (in most *STATISTICA* configurations it is limited to 1,000 characters to protect against inadvertent pasting of unwanted large amounts of data into one cell). Text in cells can be extensively formatted including different fonts and font attributes.

Input vs. Output Spreadsheets

STATISTICA offers the ability to open and use many spreadsheets at the same time, allowing you to work with several different input data files simultaneously. In addition to storing data, *STATISTICA* uses spreadsheets to display the numeric output from its analyses. Because *STATISTICA* makes no distinction in the features supported for an input spreadsheet (from which *STATISTICA* retrieves its data) and an output spreadsheet (where the results of an analysis are displayed), it is easy to use the results of one analysis as input data for further analyses.

Any spreadsheet opened from a disk file is automatically treated as an input spreadsheet, and any number of input spreadsheets can be open at a time. To avoid confusion, however, an output spreadsheet (containing the results of an analysis) is not automatically available as input data for analysis. It must first be designated as an input spreadsheet before being used for further analyses. Additionally, input spreadsheets report the number of variables and cases for that spreadsheet in the title bar. For example, if *Exp.sta (88v by 48c)* is in the title bar, it is an input spreadsheet; if *Exp.sta* is in the title bar, it is not an input spreadsheet.

To designate an output spreadsheet as an input spreadsheet, select the spreadsheet (i.e., ensure the spreadsheet has the focus), and select ***Input Spreadsheet*** from the ***Data*** menu. Now you can begin an analysis, and *STATISTICA* will use the data from

the specified input spreadsheet for the analysis. Note that if you switch back to another spreadsheet that has previously been designated as an input spreadsheet, it can still be used for analyses as well.

In a workbook, only one spreadsheet can be selected for analyses at a time, even if the workbook contains several input spreadsheets. This spreadsheet is called the Active Input spreadsheet, and its icon (in the workbook tree) is framed in red. By default, when an output spreadsheet is designated as an input spreadsheet, *STATISTICA* automatically selects it as the Active Input spreadsheet. To select another input spreadsheet for active input, select ***Use as Active Input*** from the ***Workbook*** menu or the workbook tree shortcut menu.

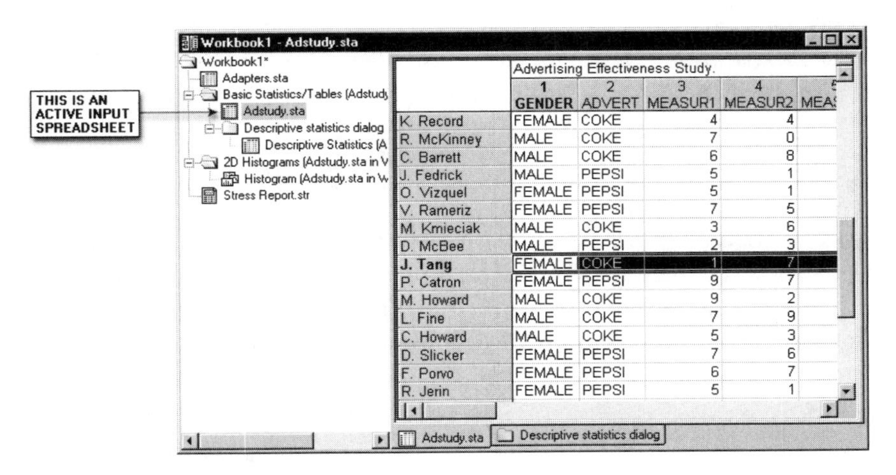

	1 GENDER	2 ADVERT	3 MEASUR1	4 MEASUR2
K. Record	FEMALE	COKE	4	4
R. McKinney	MALE	COKE	7	0
C. Barrett	MALE	COKE	6	8
J. Fedrick	MALE	PEPSI	5	1
O. Vizquel	FEMALE	PEPSI	5	1
V. Rameriz	FEMALE	PEPSI	7	5
M. Kmieciak	MALE	COKE	3	6
D. McBee	MALE	PEPSI	2	3
J. Tang	FEMALE	COKE	1	7
P. Catron	FEMALE	PEPSI	9	7
M. Howard	MALE	COKE	9	2
L. Fine	MALE	COKE	7	9
C. Howard	MALE	COKE	5	3
D. Slicker	FEMALE	PEPSI	7	6
F. Porvo	FEMALE	PEPSI	6	7
R. Jerin	FEMALE	PEPSI	5	1

It is also possible to leave a stand-alone spreadsheet open but designate it as unavailable for analysis. To do this, select the spreadsheet, and clear the ***Input Spreadsheet*** command on the ***Data*** menu. Now *STATISTICA* automatically defaults to the most recently selected input spreadsheet for analysis, ignoring all spreadsheets that are not designated as input spreadsheets.

REPORTS

Reports (briefly introduced on page 154) in *STATISTICA* offer a more traditional way of handling output (compared to workbooks) as each object (e.g., a *STATISTICA* Spreadsheet or Graph, or an Excel spreadsheet) is displayed sequentially in a word processor style document.

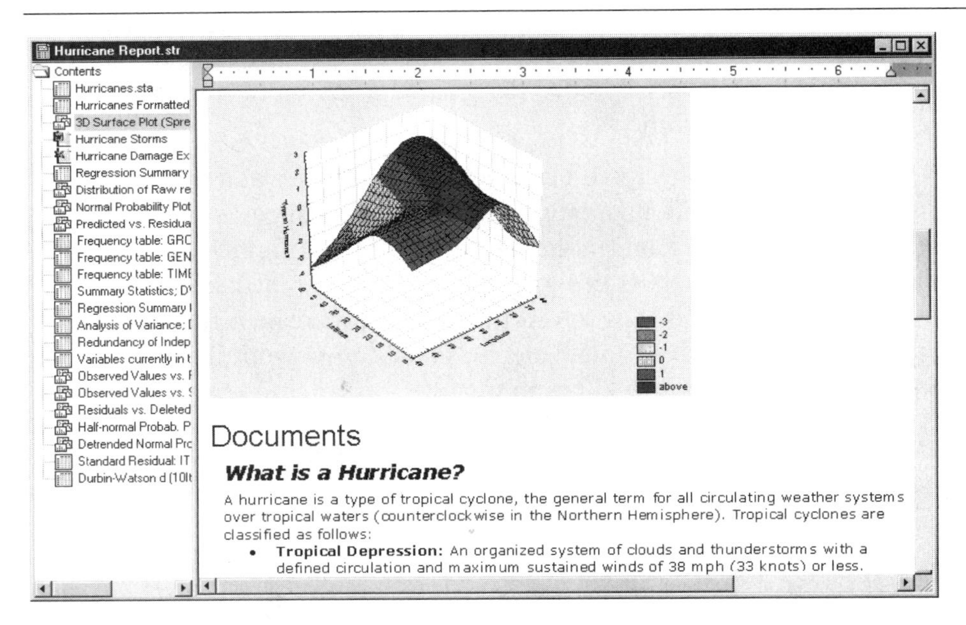

However, the technology behind this simple report offers you rich functionality. For example, like the workbook, each *STATISTICA* Report is also an ActiveX (see page 244) container where each of its objects (not only *STATISTICA* Spreadsheets and Graphs, but also any other ActiveX-compatible documents, e.g., Word documents) is active, customizable, and in-place editable. Reports are stored in the STR file format, which is a StatSoft extension of the Microsoft RTF (Rich Text Format, **.rtf*) format. STR files share the RTF formatting information and additionally they include the tree view information (which cannot be stored in the standard RTF files). Hence, report files are by default saved with the file name extension **.str*, but they can also be saved as standard RTF files (in which case the tree information will not be preserved).

The obvious advantages of this way of handling output (more traditional than the workbook) are the ability to insert notes and comments "in between" the objects as well as its support for the more traditional way of quickly scrolling through and reviewing the output to which some users may be accustomed. Also, only the report output includes and preserves a record of the supplementary information, which contains a detailed log of the options specified for the analyses (e.g., selected variables and their labels, long names, etc.) depending on the level of supplementary information requested on the ***Output Manager*** tab of the ***Options*** dialog (accessible via the ***File - Output Manager*** menu), see page 24.

The obvious drawback, however, of these traditional reports is the inherent flat structure imposed by their word processor style format, though that is what some users of certain applications may favor.

Navigating the Report Tree

The report tree displays the organization of files in the report. The files are displayed in an Explorer-style format; however, unlike workbooks that can support any number of levels, the report supports only one level of files.

You can embed any type of *STATISTICA* document in a report, including spreadsheets, graphs, and analyses. In addition to *STATISTICA* document types, you can embed other types of ActiveX/OLE objects in a report, including Excel spreadsheets, Word documents, bitmap images, and others. To edit one of these types of embedded documents, double-click on the document. The file opens in the viewer, and the report toolbar merges with the toolbar from the embedded file's native application, giving you access to all of the editing features you need.

Items in the tree are identified by the icon next to them. The spreadsheet, macro, and graph icons represent *STATISTICA* Spreadsheet, Macro, and Graph documents, respectively. The Data Miner icon represents a Data Miner workspace. All non-*STATISTICA* documents are represented by their document icons. For example, Word documents are represented by the Word icon, and Excel spreadsheet files are represented by the Excel spreadsheet icon.

The report tree can be organized and modified using drag-and-drop features as well as Clipboard procedures.

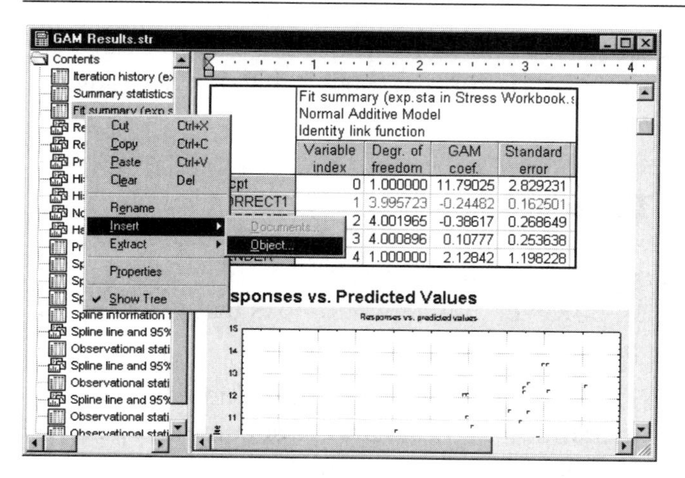

Commands for inserting, extracting, renaming, and removing items from the report tree are available from the report tree shortcut menu (accessed by right-clicking anywhere in the tree, as shown above).

GRAPHS

Graph documents represent another distinctive type of *STATISTICA* documents, and they offer rich functionality both in terms of the variety of ways in which graphs can be created in *STATISTICA* and in the selection of graph customization tools.

Similar to the other *STATISTICA* documents, graphs are ActiveX containers (see page 244), which means that they can contain a variety of compatible documents (e.g., Visio drawings, Adobe illustrations, Excel spreadsheets, etc.). *STATISTICA* Graphs are also ActiveX objects and, therefore, can be linked to or embedded into other compatible documents (e.g., Word Documents) where they can be in-place edited by simply double-clicking on them.

Graphs are discussed in more detail in Chapter 7 – *Graphs*.

MACROS (*STATISTICA* VISUAL BASIC PROGRAMS)

The industry standard *STATISTICA* Visual Basic language (integrated into *STATISTICA*) offers another (alternative) user interface to the functionality of *STATISTICA*, and it offers incomparably more than just a "supplementary application programming language" that can be used to write custom extensions. *STATISTICA* Visual Basic takes full advantage of the object model architecture of *STATISTICA* and is used to access programmatically every aspect and virtually every detail of the functionality of *STATISTICA*. Even the most complex analyses and graphs can be recorded into Visual Basic macros and later be run repeatedly or edited and used as building blocks of other applications. *STATISTICA* Visual Basic adds an arsenal of more than 13,000 new functions to the standard comprehensive syntax of Visual Basic, thus comprising one of the largest and richest development environments available.

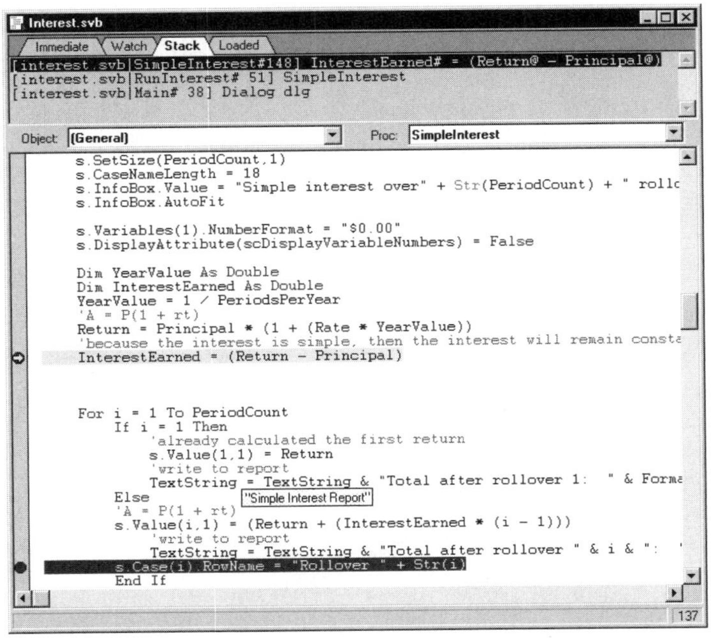

```
Interest.svb
Immediate  Watch  Stack  Loaded
[interest.svb|SimpleInterest#148] InterestEarned# = (Return@ - Principal@)
[interest.svb|RunInterest# 51] SimpleInterest
[interest.svb|Main# 38] Dialog dlg
Object: (General)    Proc: SimpleInterest
    s.SetSize(PeriodCount,1)
    s.CaseNameLength = 18
    s.InfoBox.Value = "Simple interest over" + Str(PeriodCount) + " rollc
    s.InfoBox.AutoFit

    s.Variables(1).NumberFormat = "$0.00"
    s.DisplayAttribute(scDisplayVariableNumbers) = False

    Dim YearValue As Double
    Dim InterestEarned As Double
    YearValue = 1 / PeriodsPerYear
    'A = P(1 + rt)
    Return = Principal * (1 + (Rate * YearValue))
    'because the interest is simple, then the interest will remain consta
    InterestEarned = (Return - Principal)


    For i = 1 To PeriodCount
        If i = 1 Then
            'already calculated the first return
            s.Value(1,1) = Return
            'write to report
            TextString = TextString & "Total after rollover 1:  " & Forma
        Else   "Simple Interest Report"
        'A = P(1 + rt)
        s.Value(i,1) = (Return + (InterestEarned * (i - 1)))
            'write to report
            TextString = TextString & "Total after rollover " & i & ":  '
        s.Case(i).RowName = "Rollover " + Str(i)
        End If
137
```

STATISTICA Macros can be saved in several formats, depending on how you intend to use them (see the *STATISTICA Visual Basic Primer* and the *Electronic Manual* for more information). You can also copy them to the Clipboard and paste them into other programs or documents.

STATISTICA Visual Basic is discussed in more detail in Chapter 9 (page 227).

STATISTICA PROJECTS

When performing statistical analyses and working with *STATISTICA* documents, you will often have many different windows open, and even different analyses in different stages of progress. *STATISTICA* provide a means for saving your workspace, including any analyses in progress. You can close *STATISTICA* at any point during an analysis, and when you later re-open the project, the previously opened files and in-process analyses will be restored.

To save a *STATISTICA* Project, select ***Save Project As*** from the ***File*** menu to display the ***Save STATISTICA Project*** dialog.

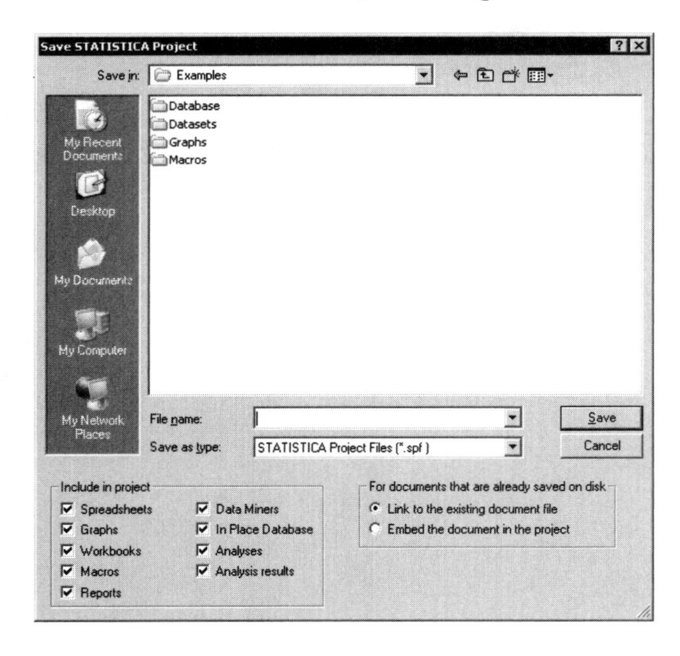

In this dialog, specify the path and file name of the *STATISTICA* Project file (with an extension of *.spf*). You can also specify what items to include in the Project. All *STATISTICA* documents types can be selected (***Spreadsheets***, ***Graphs***, ***Workbooks***, ***Macros***, ***Reports***, ***Data Miner*** projects, ***In-Place Database*** projects, ***Analyses***, and ***Analysis results***). For those *STATISTICA* documents that are already stored on disk, you have the option to either ***Link to the existing document file***, or to store a copy of the document within the *STATISTICA* Project file (***Embed the document in the project***).

In addition to *STATISTICA* documents, Project files will also save all in-progress analyses. The Project file will store the recorded scripts that are automatically created when every analysis is run. When the Project is re-opened, the scripts for the analyses are re-run against the original data and the analyses dialogs are made visible again in exactly the state they were when the Project file was saved.

Project files are a convenient way to send in-progress analysis steps and results back and forth between users if you elect to ***Embed*** the saved documents in the Project file. One user can run analyses to a certain point, and then save the Project file and pass it to another user, who can open the Project file and continue exactly where the first user stopped the analyses.

Unless you configure it otherwise, *STATISTICA* will automatically display a prompt asking if you want to save a Project file when quitting the program, and will automatically re-open the last-saved Project file when starting. Thus, *STATISTICA* makes it easy to quit for the day and start the next session right where you left off.

Note that a project is a state of an instance of *STATISTICA*. Thus, projects are not like other documents in that you cannot open more than one project in a single instance of *STATISTICA*. A different (second) project can be opened in a second instance of *STATISTICA*.

DATA MINER RECIPE

The *Data Miner Recipe* provides the solution that maps the steps of the data mining workflow into a results-oriented user interface. From data cleaning to model validation, *Data Miner Recipe* guides your analysis from start to finish so that you can get actionable results and answers quickly. At the same time, the *STATISTICA Data Miner Recipe* still applies the most comprehensive collection of data mining algorithms in a single package, without requiring the user to know the details of those algorithms.

STATISTICA Data Miner contains the largest collection of data mining methods and algorithms in a single package or library. In most general terms, these algorithms borrow insights and methodologies from various domains such as statistics, engineering, artificial intelligence, cognitive science, etc., to “learn” patterns from data that can be used to make predictions (about insurance or credit risk, process or product quality, equipment failure, medical diagnoses, and so on). The *STATISTICA Electronic Manual* and the on-line *Electronic Statistics Textbook* provide detailed introductions to the various methods and techniques that are usually summarily described as “data mining.”

In practice, specific domains and types of data are best analyzed using particular types of methods and algorithms. For example, the data mining techniques that work best for modeling insurance loss data will likely be (and are) different from those that work best for predicting emissions from a furnace. However, there *is* a typical workflow – from the definition of the data and analysis problem through sampling, model building, and evaluation – that is applicable to all predictive data mining.

The *Data Miner Recipe* user interface and project metaphor enable those without extensive experience with data mining tools to move very quickly from the definition of a problem to tangible and actionable results.

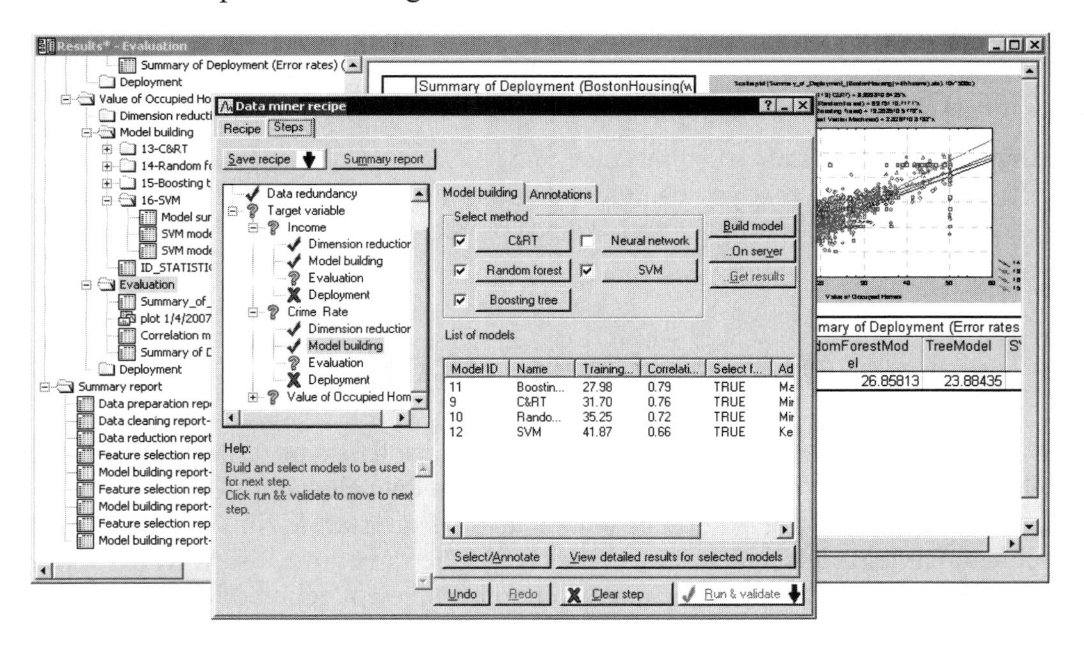

In this approach, you simply follow a recipe-like user interface to complete the necessary steps to move to a solution. In fact, most of these steps are entirely automated so that the only required input is to define the data and variables for the analyses, while the program automatically does the rest – determines learning and testing samples, performs feature selection, tries various data mining algorithms and methods, and automatically evaluates results to select the best data mining model. These computations and analyses can be performed with either the desktop *STATISTICA Data Miner* software or, if available, on the *WebSTATISTICA Data Miner* server.

Data Miner Project Files

When you save a data miner project (*Data Miner Recipe*) at any stage of completion, actually two separate files are created:

- A *Data Miner Recipe* file with the file name extension *.dmrproj*
- A *STATISTICA Workbook* file by the same name, but with the file name extension *.stw*, containing results and detailed information for each step of the recipe

It is important that *both* files reside in the same file directory. So, if you want to copy a *Data Miner Recipe* project called *MyDataMinerProject* to a new file directory, email it to a colleague, or check it into the *STATISTICA Document Management System*, then both files – *MyDataMinerProject.dmrproj* and *MyDataMinerProject.stw* – must be copied to the new destination.

Here are some additional details about these two files.

Data Miner Recipe file (.dmrproj). The *Data Miner Recipe* is an *XML* (extensible markup language) format file that contains all information regarding users' choices (or choices automatically made by the program), including:

- Data file information (or data connection information)
- Variable selections and variable metadata (e.g., defining continuous and categorical predictors and outcomes)
- Choices about data preprocessing steps (e.g., missing data handling, filtering of duplicate records, transformations, etc.)
- Final variable selections based on the application of feature selection algorithms
- Results from model building and final evaluation and choices of models

- All information necessary to deploy predictive models and to predict new cases (e.g., to score databases, compute component scores, inferred sensor values, predicted risk or failure probabilities, etc.)

Therefore, when deploying *Data Miner Recipes* to the *STATISTICA Enterprise* software to automatically compute predicted values in an enterprise application (automated credit scoring, multivariate control charting and failure analysis, etc.), all information necessary to compute predicted values, classifications, or classification probabilities (e.g., probability of default, loss) is contained inside these XML format files.

Data Miner Recipe Workbook file (*.stw*). These files contain detailed information describing the results for each step.

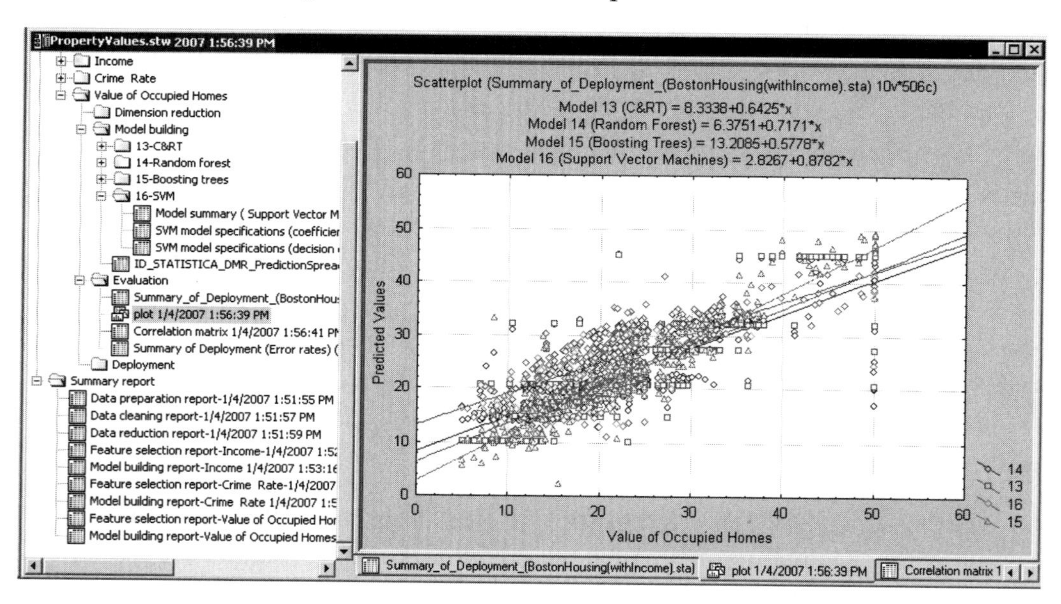

The results stored in this workbook provide complete documentation for the computations and analyses performed as the *Data Miner Recipe* was (or is in the process of being) completed. Therefore, if the data mining analyses are performed in a regulated (e.g., FDA, ISO, etc.) environment, or if data mining is part of an organization's mission critical activities performed under the guidance and in compliance with specific standard operating procedures (SOPs), then it is usually recommended that this file be stored in the *STATISTICA Document Management System* along with the *Data Miner Recipe* project file (*.dmrproj*).

7

CHAPTER

GRAPHS

CHAPTER

GRAPHS

OVERVIEW

The most common application of graphs is to efficiently present and communicate information (typically, numerical data). However, graphical techniques also provide powerful analytical tools for the exploration of data and verification of hypotheses.

A broad selection of graphics options. *STATISTICA* includes a comprehensive selection of graphical methods for both data analysis and the presentation of results. All graphs in *STATISTICA* include a broad selection of built-in, interactive analytic techniques and extensive customization tools that enable you to interactively control virtually all aspects of the display. Also, flexible graphics management facilities are available that are used to integrate various graphical displays and to build dynamic links between applications (e.g., using OLE-Object Linking and Embedding).

Comprehensive support for Visual Basic and other languages. *STATISTICA* graphical options can also be accessed programmatically (using built-in *STATISTICA* Visual Basic or other compatible languages), which creates practically unlimited possibilities for producing highly customized graphical displays. These custom graphs can later be permanently added to *STATISTICA*'s user interface (e.g., assigned to buttons on toolbars or added to the menus).

General categories of graphs. The *STATISTICA* system offers a variety of methods in which graphs can be requested or defined. These methods (constituting broad categories of graphs, such as input data, block data, and specialized) are reviewed in *General Categories of Graphs* on page 204; they complement each other, providing a high level of integration between numbers (such as raw data,

intermediate results, or final results) and graphical displays. For example, specialized graphs can be requested as part of the automatic output from statistical procedures, but they can also be requested via integrated tools to visualize virtually any combination of numbers (and/or labels) that are displayed or generated by *STATISTICA*.

CUSTOMIZATION OF GRAPHS

Interactive graph customization. The customization options in *STATISTICA* graphics include hundreds of features and tools that can be used to adjust every detail of the display and associated data processing. However, these options are arranged in a hierarchical manner, so those used most often are accessible directly via shortcuts by double-clicking or right-clicking on the respective element of the graph.

Permanent settings and automation options. The initial (default) settings of all of these features can be easily adjusted so that even the default appearance and behavior of *STATISTICA* graphs will match your specific needs and/or will require very little intervention on your part. There are at least four different ways to make these adjustments:

1. ***Options* dialog.** Perhaps the most straightforward way to adjust the default appearance of graphs is by using the ***Graphs 1*** and ***Graphs 2*** tabs of the ***Options*** dialog (accessible via the ***Tools - Options*** menu). Most commonly used settings can be easily adjusted here, and the results will be reflected in the default styles (see number 2 below) that will be used by the system and as such, they will be automatically saved in the *STATISTICA* configuration file (e.g., different settings can be used for different projects). For further details, see the documentation for the ***Configuration Manager*** tab of the ***Options*** dialog in the *Electronic Manual*.

2. **Graph style system.** All of the numerous features that affect the appearance of the graph (from as elementary as the color of the font in the footnote to as general as the global features of the graph document) can be saved as individual "styles." These styles can be given custom names and later be reapplied using simple shortcuts (such as pressing a specific key combination

or clicking a button on a custom toolbar). An intelligent system internally manages these thousands of styles and their combinations in *STATISTICA* and helps you achieve your customization objectives with a minimum amount of effort. All user-defined or modified styles will be saved automatically in the *STATISTICA* configuration file (e.g., different sets or systems of styles can be used for different projects). For further details, see the documentation for the ***Configuration Manager*** tab of the **Options** dialog in the *Electronic Manual*.

3. **User-defined graphs.** New types of graphs can be defined in a variety of ways and can be added to the menus, dialogs, or toolbars. If a custom graph that you intend to use repeatedly is not built "from scratch" but is based on one of the ***Graphs*** menu graphs and is produced by some combination of the existing graph customization options, then adding it to the ***Graphs*** menu as a new type of graph is as simple as clicking the ***Add As User-defined Graph to Menu*** button on the ***Options 2*** tab of the graph specification dialog. All user-defined graph specifications will be saved automatically in the *STATISTICA* configuration file (e.g., different sets of custom graphs can be used for different projects). For further details, see the documentation for the ***Configuration Manager*** tab of the ***Options*** dialog in the *Electronic Manual*.

4. ***STATISTICA* Visual Basic.** Finally, note that there are no limits to how "deeply customized" your *STATISTICA* custom graphs can be, because *STATISTICA* Visual Basic (with all its powerful custom drawing tools as well as the *STATISTICA*-based library of graphics procedures) can be used to produce virtually any graphics or multimedia output supported by the contemporary computer hardware. Those custom developed displays or multimedia output can be assigned to *STATISTICA* toolbars, menus, or dialogs and become a permanent part of "your" *STATISTICA* application.

See the *Electronic Manual* for further details on these graph customization methods.

The *Electronic Manual* also contains topics devoted to specific categories of graphs, includes conceptual overviews and examples of typical applications, and discusses distinctive functional properties of the respective types of graphs.

The default settings of most graphs offered in *STATISTICA* follow the established conventions that are either explicitly described in the literature on statistical and technical graphing, or they represent standards that are commonly accepted by major scientific journals (e.g., *SCIENCE*). However, practically all default settings of *STATISTICA* can be customized to meet specific requirements of unusual

applications (see page 198). *STATISTICA*'s graphics facilities were designed to play the role of flexible tools, capable of producing effects that go far beyond established patterns and templates. Moreover, these tools can be customized and new tools can be designed, and they can be added to toolbars or menus for repeated use (see page 198).

In addition to a comprehensive selection of standard statistical and technical graphs, *STATISTICA* includes numerous unique types of graphs and graph customization facilities. While StatSoft statisticians designed most of them, it is important to say that *STATISTICA* users have played a significant role in their creation. In fact, the final selection of graphics options included in *STATISTICA* is the result of input from thousands of users who provided their comments in response to StatSoft's inquiries. Many unique facilities of *STATISTICA* Graphs (e.g., the multiple subset selection facility and the on-line categorization options) were introduced in response to users' ideas and requests. We at StatSoft are very grateful for the input from our users.

As mentioned previously (and discussed in detail on page 204) there are various methods to request *STATISTICA* Graphs. You could say that these methods represent different types of "interfaces" between numbers and graphs.

For example, the numbers represented in a pie chart can simply depict values of a spreadsheet column (e.g., variable *Sales*) in the consecutive cases of the spreadsheet (e.g., cases labeled: *Year 2002, Year 2003, Year 2004, ...*, etc.). The numbers in a similar pie chart, however, can also represent results of some calculations. For example, the slices of the pie can represent relative frequencies of observations that belong to certain categories calculated by one of the histogram or frequency categorization procedures (e.g., numbers of years when the *Sales* were below $10 million, between $10 and $20 million, and above $20 million).

Regardless of the method that was used to create a graph (i.e., regardless of where the numbers represented in the graph were obtained or how they were calculated), all *STATISTICA* Graph customization and multigraphics management facilities can be used to change the appearance of the graph or integrate it with other graphs or documents.

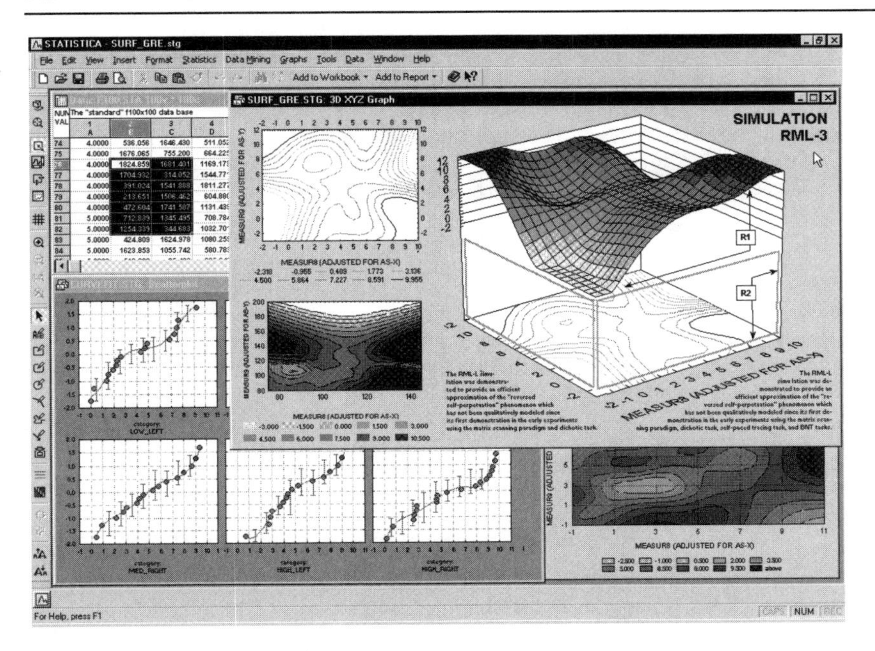

Also, all integrated analytic facilities that are accessible from within graphs in *STATISTICA* (such as function fitting, smoothing, rotation, brushing, analytical zooming, etc.) are available and can be applied to the graph regardless of the source of the numbers in the graph or the method that was used to create it.

The graph editing facilities offered in *STATISTICA* enable you to create not only highly customized scientific and technical publication-ready displays:

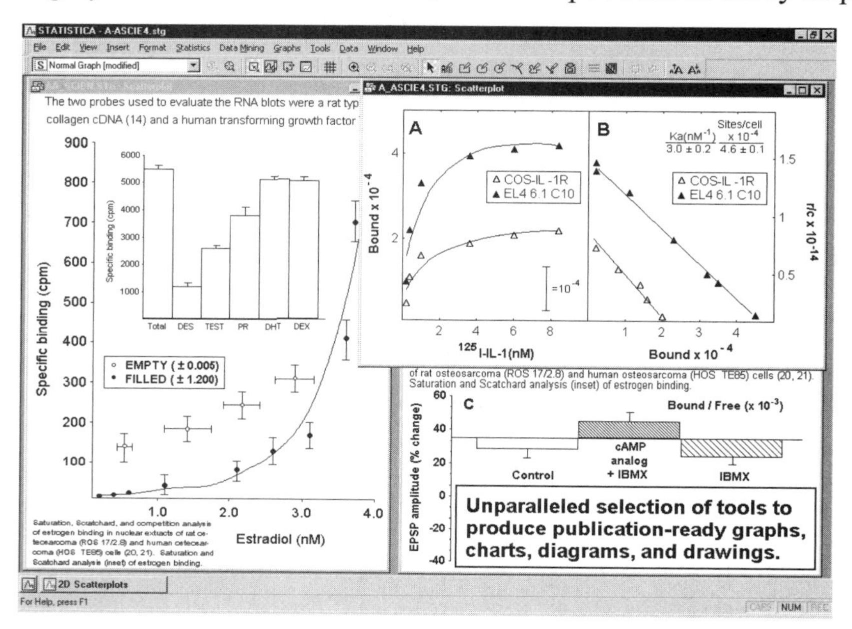

and precise drawings:

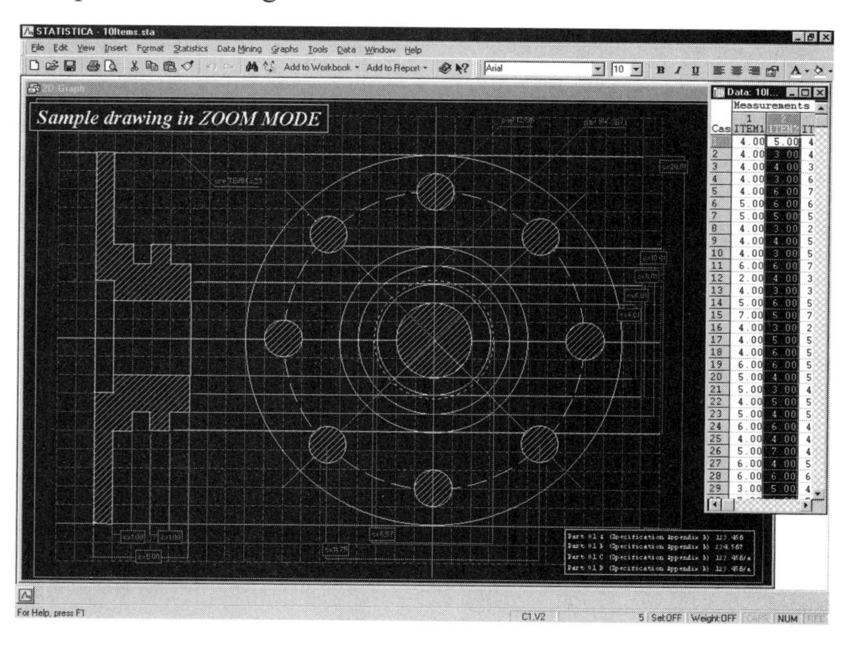

but also presentation-quality diagrams, posters, business charts, and other displays:

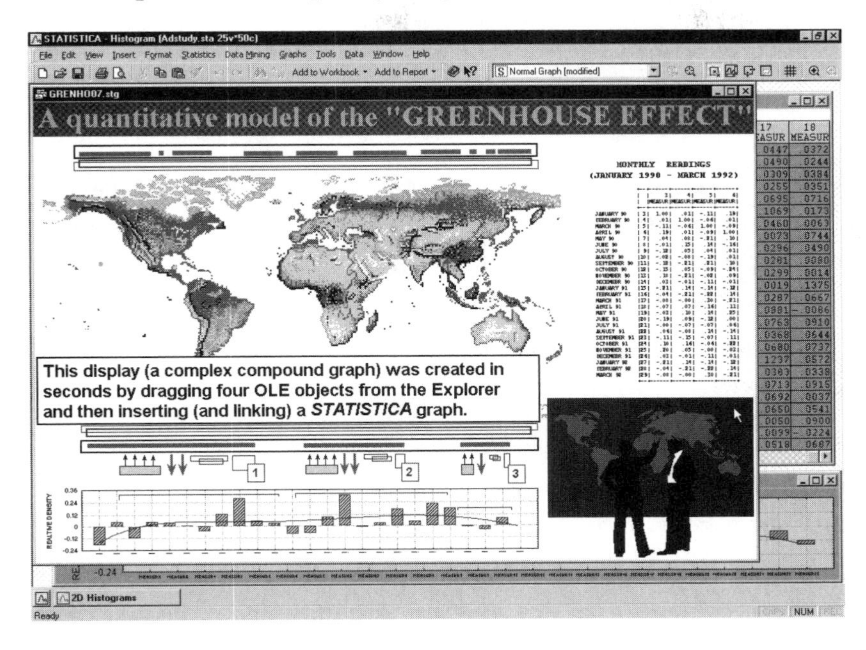

that are designed to communicate information in an effective and attractive manner.

Graphs that are saved into files or that in any other way have been temporarily detached from the *STATISTICA* application (e.g., copied to the Clipboard or linked to a document in another application) are complete "objects" (technically speaking, ActiveX objects, see page 244) that contain not only all customization features and other embedded objects, but also all data that are necessary to continue editing all aspects of the display or the analysis of its contents (fitting, smoothing, etc.).

Because *STATISTICA* Graphs are ActiveX objects, they can easily be linked to or embedded in other compatible documents (e.g., Excel or Word documents), where they can be in-place edited by double-clicking on them. *STATISTICA* Graphs are also ActiveX containers and, therefore, can contain a wide variety of embedded or linked documents such as Visio drawings, Adobe illustrations, Excel spreadsheets, or Word documents. Moreover, *STATISTICA* supports hierarchies of embedded objects up to four levels, which means that it can manage "documents containing documents, containing documents, which contain documents."

GENERAL CATEGORIES OF GRAPHS

In addition to the specialized statistical graphs that are available from the output dialogs in all statistical procedures (see page 215), there are two general categories or classes of graphs both accessible from the ***Graphs*** menu, ***Graphs*** toolbar, shortcut menus, and the *STATISTICA* Start button menu:

- Input data graphs (***Graphs of Input Data***, see page 205, and ***Graphs*** menu graphs, see page 209) and
- ***Graphs of Block Data*** (see page 208).

The most important difference between these two general categories lies in the data that the graph types utilize for generating plots.

Input data graphs. ***Graphs of Input Data*** and their expanded version in the ***Graphs*** menu produce statistical summaries or other representations of the raw data in the current input data spreadsheet (typically for all the variable(s), or for subsets if case selection conditions are used). Note that if graphs of this general category are produced using a shortcut menu from within a spreadsheet of results that does not contain the actual data (e.g., a correlation matrix), *STATISTICA* will still reach to the respective input (raw) data in order to produce the graph (e.g., a scatterplot of the variables identified by the selected cell in the correlation matrix from which the shortcut menu was opened).

Graphs of Block Data. ***Graphs of Block Data***, on the other hand, are entirely independent of the concept of "input data" or "data file." They provide a general tool to visualize numeric values in the currently selected block of any spreadsheet (which can contain values from custom defined subsets of numerical output or arbitrarily selected subsets of raw data).

Common features of the two categories of graphs. Note that these two general categories of graphs offer the same customization options and the same selection of types of graphs. For example, you can create the same, highly specialized categorized ternary graph from the input (raw) data set, and from a custom defined block of values representing results of a particular test.

These two general categories of graphs will be briefly discussed in the next two sections, followed by a section on the ***Graphs*** menu, which contains an exhaustive selection of all graphs from the first category (input data graphs, often referred to as ***Graphs*** menu graphs), as well as access to ***Graphs of Block Data*** and other options.

GRAPHS OF INPUT DATA

The ***Graphs of Input Data*** command is available from the shortcut menu of all spreadsheets, and it offers quick and simplified access to the most commonly used types of graphs based on the current input data set.

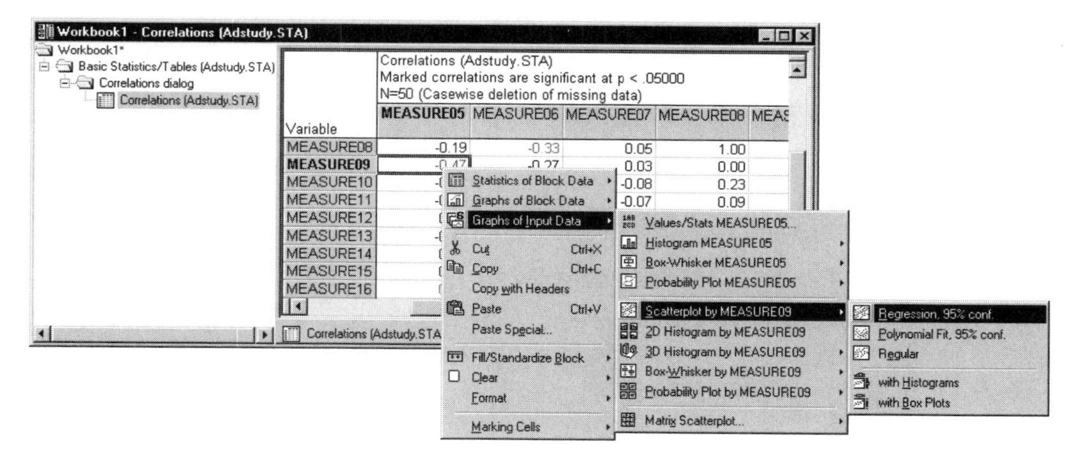

Note that all these graphs are also available from the ***Graphs*** menu, from the *STATISTICA* Start button menu on the status bar, or by clicking the ***Graphs Gallery*** button on any graph specification dialog. ***Graphs of Input Data*** do not offer as many options as the corresponding ***Graphs*** menu graphs; however, they are quicker to select because unlike ***Graphs*** menu graphs:

- ***Graphs of Input Data*** can be called directly from the spreadsheet shortcut menus,
- ***Graphs of Input Data*** do not require you to select variables (the variable selection is determined by the current cursor position within a spreadsheet), and
- ***Graphs of Input Data*** do not require you to select options from any intermediate dialogs (default formats of the respective graphs are produced).

Graphs of Input Data process data directly from the current input data file, and they take their cues as to which variables to use from the current cursor position (in any type of spreadsheet).

For example, if you right-click a single correlation in a results spreadsheet and create a ***Scatterplot by...*** graph, *STATISTICA* generates a 2D scatterplot using the original raw values of the two variables represented by that correlation (see the *Introductory Example* on page 11 for a more detailed example).

Although the most convenient (and you could say most logical) way to select ***Graphs of Input Data*** is via the spreadsheet shortcut menu, you can also select them from the ***Graphs*** menu or the *STATISTICA* Start button menu. Either method will display a submenu from which you can choose one of the statistical graphs applicable to the current variable (i.e., to the variable indicated by the current cursor position in the spreadsheet).

If the spreadsheet has a matrix format or a format where a cursor position indicates not one but two variables (as in the illustration showing a correlation matrix, below), then predefined bivariate graphs for the specified pair of variables will be directly available from the ***Graphs of Input Data*** menu.

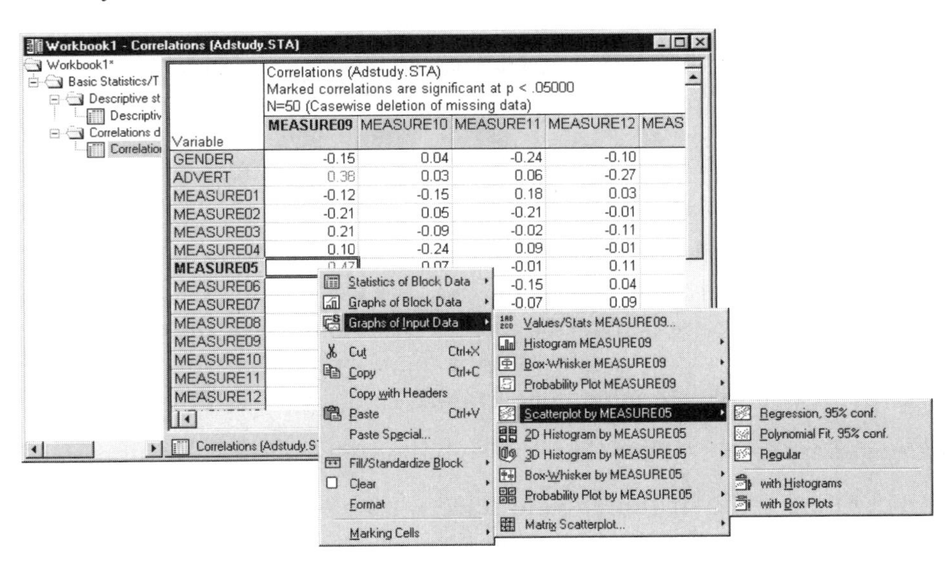

Otherwise, i.e., when the current cursor position indicates only one variable as in a table of descriptive statistics (as shown in the next illustration), and if you select any of the bivariate graphs in the menu, *STATISTICA* will prompt you to select the

second variable. For example, if you select ***Scatterplot by***, the ***Select second variable*** dialog will be displayed, where you specify by which variable ***Measure05*** is going to be plotted.

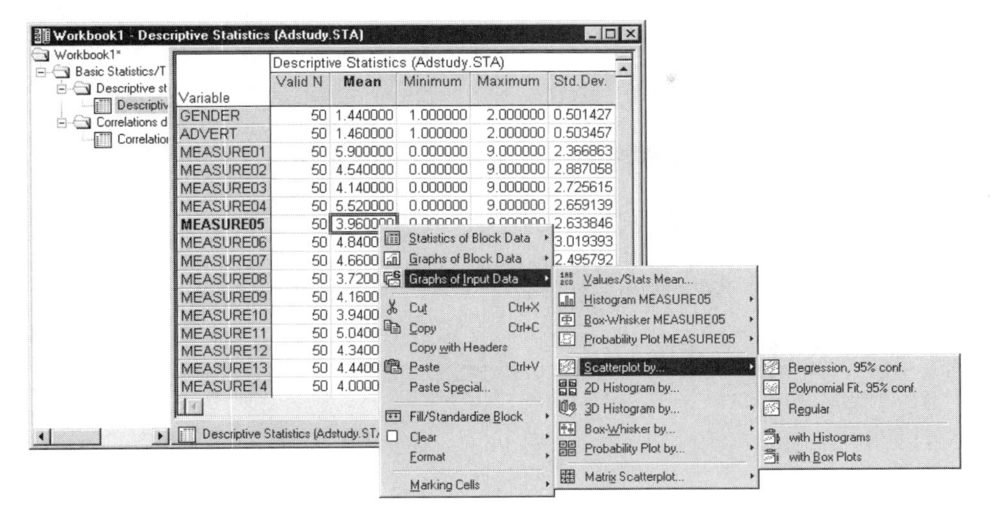

If more than one variable is indicated by a highlighted section (i.e., when a block is selected), then the ***Graphs of Input Data*** menu will apply to the first selected variable.

When generating ***Graphs of Input Data***, *STATISTICA* takes into account the current case selection and weighting conditions for the variables that are being plotted. Note, however, that the case selection or weighting conditions need to be specified for the current spreadsheet (i.e., using the ***Tools - Selection Conditions - Edit*** and the ***Tools - Weight*** menu commands) and not only "locally" for an analysis (i.e., selected from the respective analysis/graph specification dialogs using the [SELECT CASES] and [W] buttons). The latter conditions will be ignored by the ***Graphs of Input Data***. For more information on specific types of ***Graphs of Input Data***, see the *Electronic Manual*.

GRAPHS OF BLOCK DATA

Unlike ***Graphs of Input Data***, ***Graphs of Block Data*** use the currently selected (continuous) block of data in the active spreadsheet to specify input data for the graph.

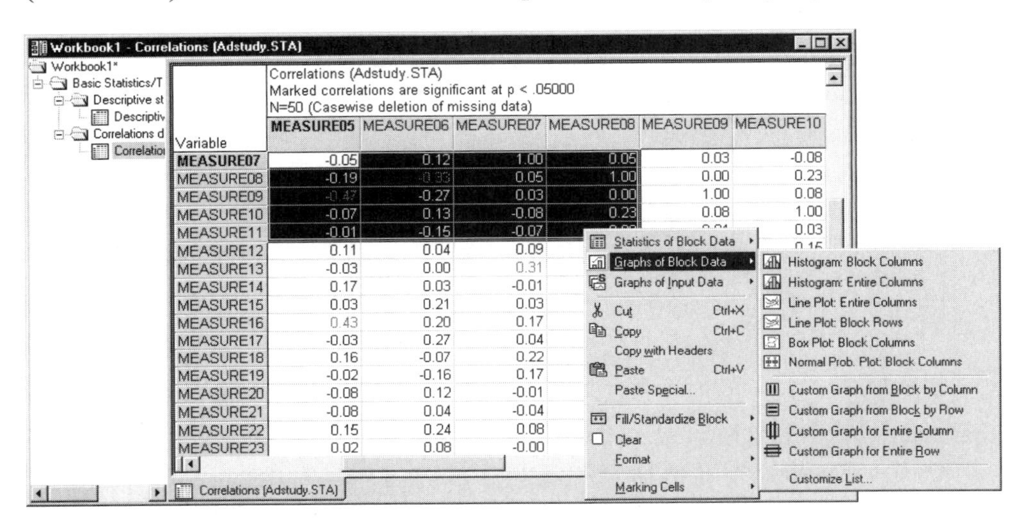

Note that these graphs are entirely independent from the concept of "input data." They process values (numbers) from whatever is currently selected in the block and ignore the "meaning" of those numbers (e.g., the numbers can be raw data or values of correlation coefficients). These graphs offer an effective means of visualizing, exploring, and efficiently summarizing numeric output from analyses displayed in results spreadsheets (e.g., histograms of Monte Carlo output scores in the ***SEPATH*** module, or a box plot of aggregated means from a multivariate multiple classification table in the ***ANOVA*** module).

Although the most convenient (and one can say most logical) way to select ***Graphs of Block Data*** is via the shortcut menu associated with the block selected in a spreadsheet, ***Graphs of Block Data*** are also available from the ***Graphs*** menu or the *STATISTICA* Start button menu. When creating ***Graphs of Block Data***, you can select from default graphs (e.g., ***Histogram: Block Columns*** or ***Line Plot: Block Rows***), or you can create your own custom graphs for either the selected cells in the rows or columns or of all cells in the selected rows or columns (i.e., going beyond the values that are selected in the block).

Default graphs. Using the default graphs (the first six commands on the ***Graphs of Block Data*** submenu, shown in the illustration above), you can create specified graphs with a single click. For specific information on each default graph, refer to the *Electronic Manual*.

Custom graphs. Select one of the four ***Custom Graph*** commands to display the ***Select Graph*** dialog, which provides a variety of options for creating customized graphs.

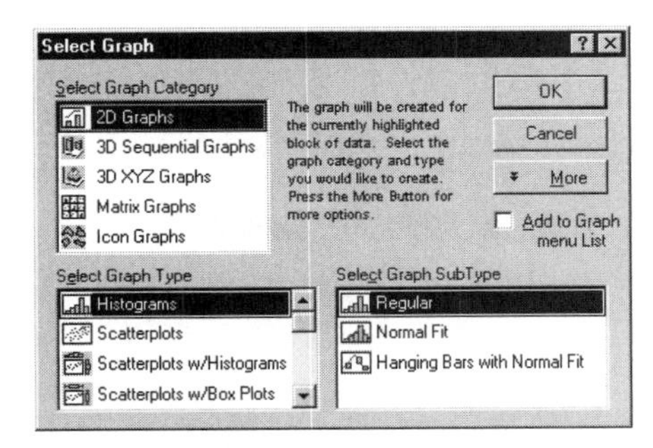

For specific information on custom graphs, refer to the *Electronic Manual*.

Customizing graphs. As with most features of *STATISTICA*, ***Graphs of Block Data*** are fully customizable. Select ***Customize List*** from the ***Graphs of Block Data*** menu to display the ***Customize Graph Menu*** dialog, which provides options to remove, rename, or edit the currently listed graphs as well as to add new (user-defined) graphs to the ***Graphs of Block Data*** menu.

For example, if you want to include a normal fit on the histograms created using ***Histogram: Block Columns***, select ***Histogram: Block Columns*** in the ***Customize Graph Menu*** dialog, click the ***Edit*** button, and switch the ***Graph SubType*** to ***Normal Fit***. All subsequently created ***Histogram: Block Columns*** plots will include a normal fit to the data.

GRAPHS MENU GRAPHS

The ***Graphs*** menu provides a complete selection of all statistical graphs available in *STATISTICA*. These commands are available from not only the ***Graphs*** menu, but also the *STATISTICA* Start button menu, and offer hundreds of types of graphical representations and analytic summaries of data.

Note that, unlike ***Graphs of Block Data*** (which are also included on this menu in order to offer a full complement of all graphical options accessible from a single control), all other graph types from the ***Graphs*** menu are not limited to the values in the current output spreadsheet. Instead, they process data directly from the current input spreadsheet, in the same way the (previously discussed) ***Graphs of Input Data*** do. They represent either standard methods to graphically summarize raw data (e.g., various scatterplots, histograms, or plots of central tendencies such as medians) or standard graphical analytic techniques (e.g., categorized normal probability plots, detrended probability plots, or plots of confidence intervals of regression lines). When generating these graphs, *STATISTICA* takes into account the current case selection and weighting conditions for the variables selected to be plotted.

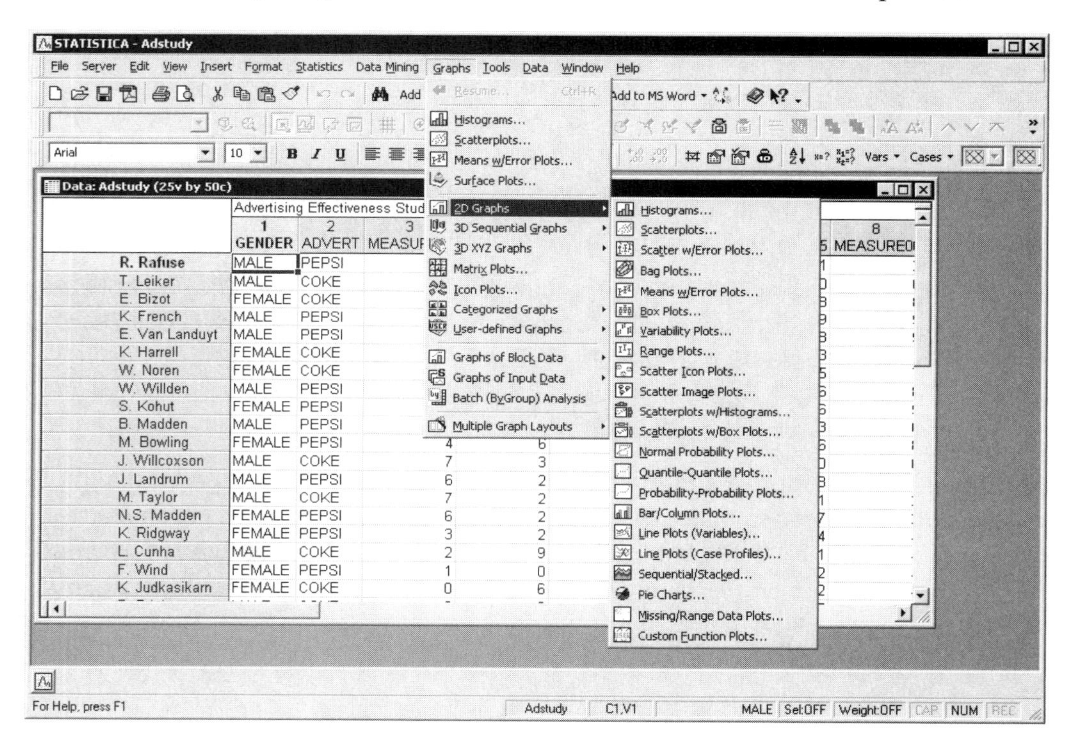

Graphs menu graphs include ***2D Graphs***, ***3D Sequential Graphs***, ***3D XYZ Graphs***, ***Matrix Plots***, ***Icon Plots***, ***Categorized Graphs***, and ***User-Defined Graphs***. Note that the top portion of the ***Graphs*** menu includes the most commonly used types of graphs (***Histograms***, ***Scatterplots***, ***Mean/Error Plots***, etc.), and the lower portion contains a comprehensive list of all graph types. Like all menus in *STATISTICA*, it

can be easily customized (use the ***Menu*** tab of the ***Customize*** dialog, accessible from the ***Tools - Customize*** menu) to position the most commonly used options in the most convenient locations. See also, *Types of Graphs Menu Graphs* in the *Electronic Manual*.

GRAPH BRUSHING AND CASE STATES

Graphs that are created from the ***Graphs*** menu are highly interactive with the spreadsheet from which they were created. You can identify and select points in the graph and specify that they are to be highlighted in the source spreadsheet, and vice versa. In addition to selecting points in graphs and spreadsheets, you can identify properties of a case in a spreadsheet that will be used when the graph is created from that data. These properties include the point marker style and color, and whether the point is to be excluded from the graph and/or fit calculations.

To start brushing within a graph, click the brushing toolbar button or right-click in the background of a graph and select ***Show Brushing*** from the shortcut menu to display the ***Brushing*** dialog, which is shown in the illustration.

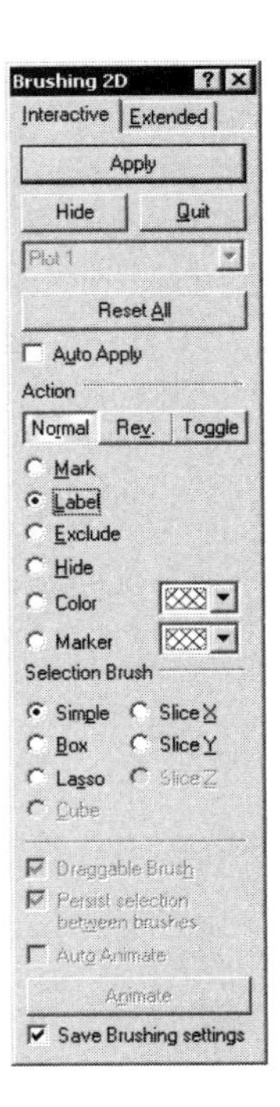

With the default ***Selection Brush***, which is ***Simple***, you can draw a rectangle on the graph to select the points contained in the rectangle. The following illustration demonstrates this for the example data set *Adstudy.sta*, with a 2D scatterplot of *MEASURE01* by *MEASURE02*. Note that the upper-left three points have been selected by the brushing tool, which highlights the points in the graph as well as the corresponding cases in the spreadsheet from which the graph was created.

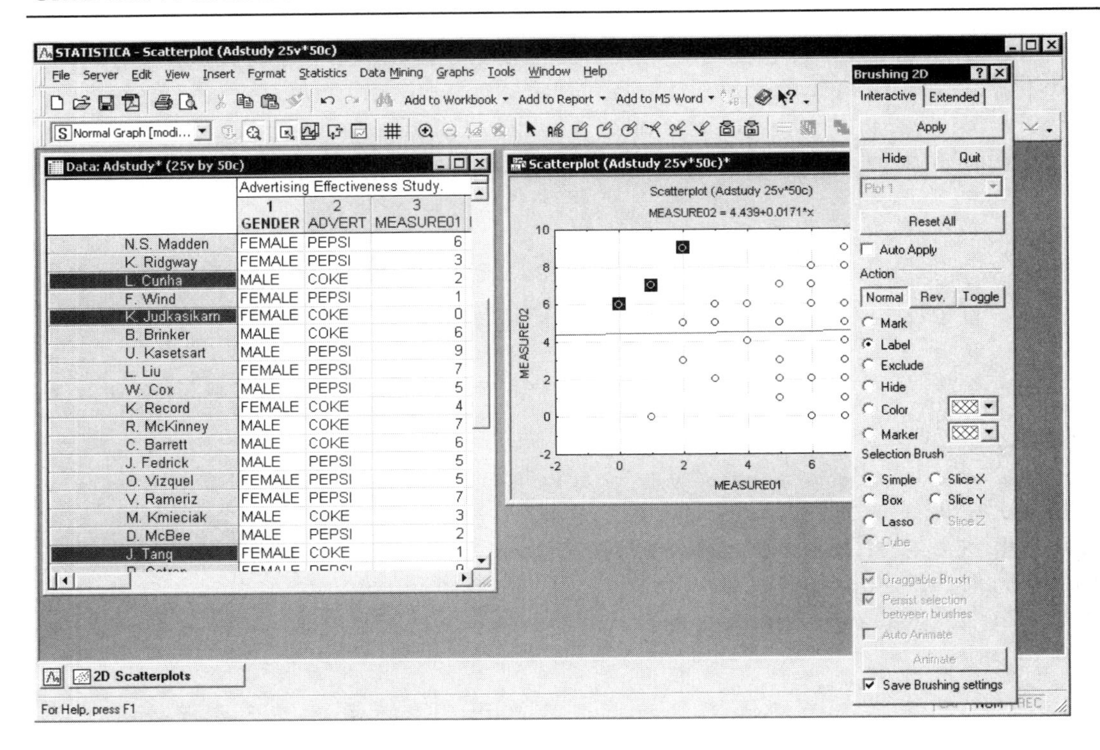

Alternatively, instead of using the ***Brushing*** facilities, you can select cases in the spreadsheet (click on the far-left side of the case name) and the corresponding points will be marked in the graph, as shown in the following illustration, where the first five cases in the *Adstudy.sta* spreadsheet have been selected.

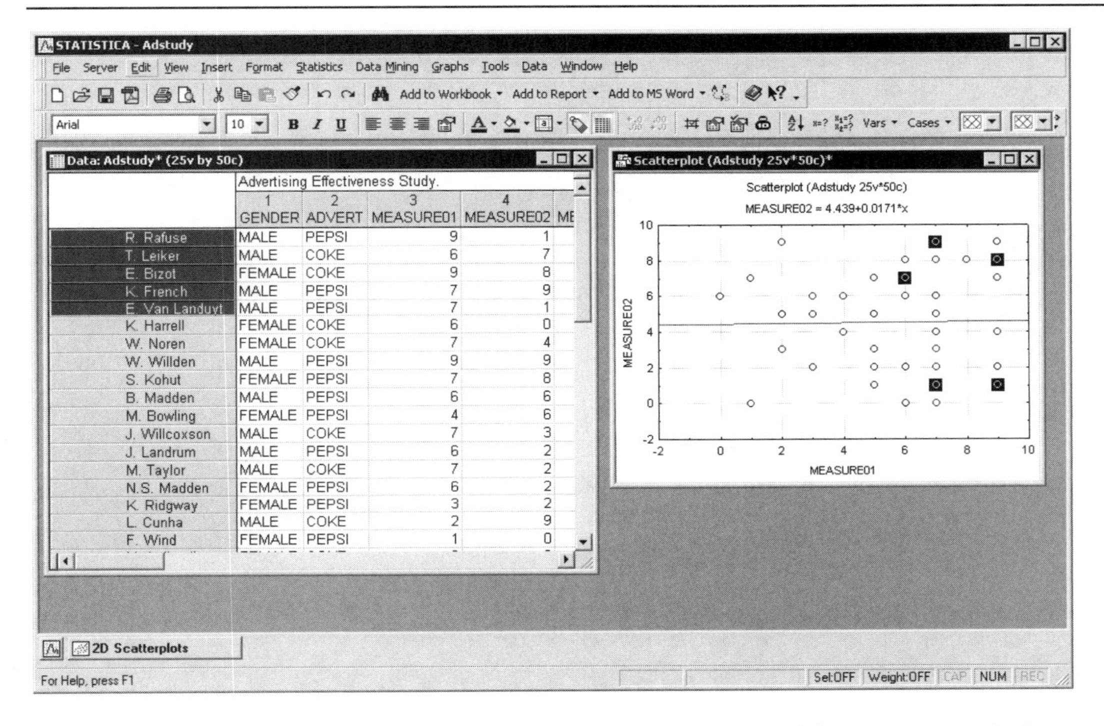

You can specify spreadsheet case states from either a spreadsheet or a graph. In a *STATISTICA* Spreadsheet, right-click on a case name to display the shortcut menu, which contains commands including ***Excluded***, ***Hidden***, ***Label***, ***Marked Points***, and ***Case States***. Similar commands are available from the shortcut menu displayed when you right-click on the points in a graph. The graph will use these options when displaying the points represented by this case. For example, if you select ***Label***, the corresponding points will be labeled, as shown in the next illustration. Note that the spreadsheet cases are marked with a case state icon to indicate the cases are labeled:

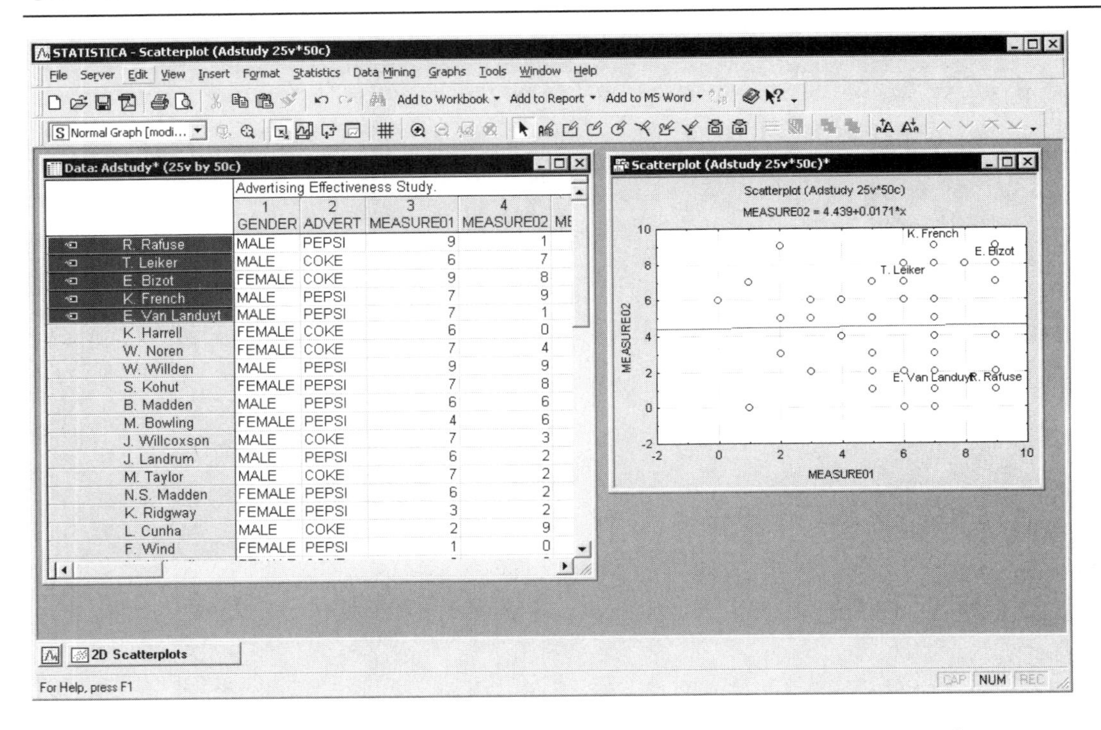

Right-click on a case name and from the shortcut menu - ***Case States*** submenu, select ***Edit Case States*** to change the case marker and/or color.

Note that the selection of points is available for graph types other than Scatterplots. For histograms, brushing/selecting a histogram bar will select the corresponding points to that bar in the spreadsheet. The same is true of the boxes in box plots.

Using case states and brushing and selecting points is particularly useful with the ***Hidden*** and ***Excluded*** case states options. From the ***Data - Cases - Case States*** submenu, select ***Hidden*** to mark a case as hidden, i.e., the case will not be visible in graphs, but will be used in analyses. You can also right-click on a case name and access this option from the shortcut menu. Select ***Excluded*** from either of these menus to mark a case as excluded, i.e., the case will not be used in the computations; however, the case will be displayed in most graph types. The case point marker is displayed but the case is removed from computations. To remove the effect of a point from a graph, select both options. The ***Excluded*** case state also works in conjunction with spreadsheet selection conditions; any case that has the ***Excluded*** case state set will be treated as if the case were excluded by selection

conditions. Therefore, using graph brushing and case states is a convenient tool to interactively remove outliers, and then re-run analyses with the points removed.

OTHER SPECIALIZED GRAPHS

In addition to the standard selection of ***Graphs of Input Data***, ***Graphs of Block Data***, and ***Graphs*** menu graphs (see above), other specialized statistical graphs that are related to a type of analysis (e.g., cluster analysis results) are accessible directly from results dialogs (i.e., the dialogs that contain output options from the current analysis).

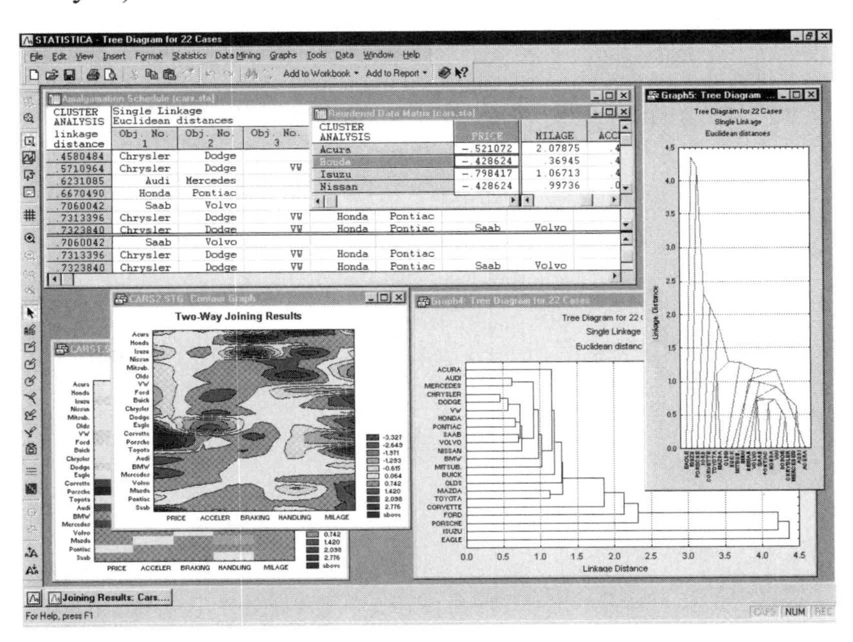

The specialized graphs are described in the documentation for the analyses from which they can be produced; thus, for specific information, refer to the respective sections of the *Electronic Manual*.

CREATING GRAPHS VIA *STATISTICA* VISUAL BASIC

STATISTICA graphical options can also be accessed programmatically using the built-in *STATISTICA* Visual Basic or other compatible languages. Therefore, there are no limits to how "deeply customized" your *STATISTICA* custom graphs can be, because *STATISTICA* Visual Basic (with all its powerful custom drawing tools as well as the *STATISTICA*-based library of graphics procedures) can be used to produce virtually any graphics or multimedia output supported by the contemporary computer hardware. Those custom developed displays or multimedia output can be assigned to *STATISTICA* toolbars, menus, or dialogs and become a permanent part of "your" *STATISTICA* application.

An application written in *STATISTICA* Visual Basic can operate on graphs in three ways:

- Make a new graph and then modify, print, or save it;
- Access an existing graph window and then modify it;
- Open an existing graph file and then modify, print, or save it.

Every graph available in *STATISTICA* can be produced by *STATISTICA* Visual Basic and then customized using *STATISTICA* procedures or general options offered in this comprehensive language.

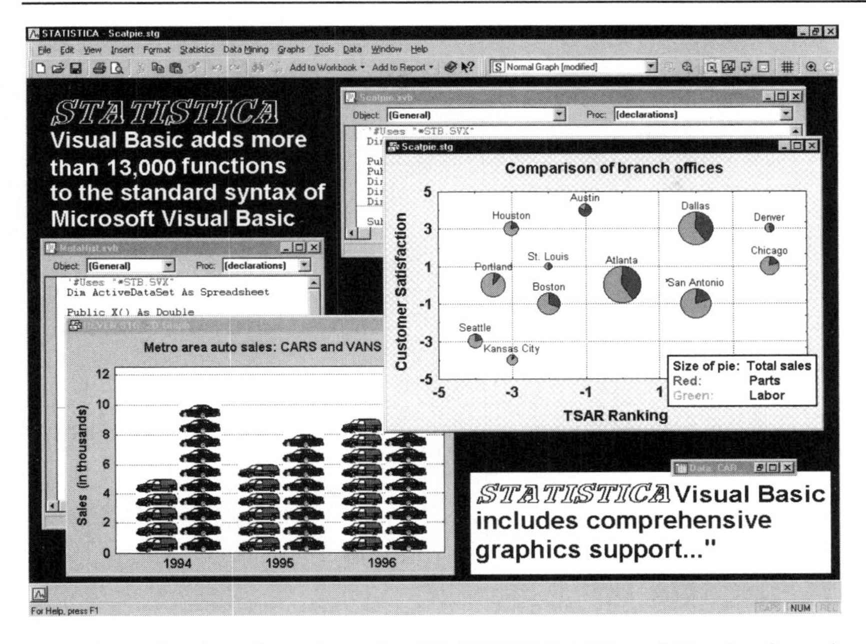

As with all other functions in *STATISTICA* Visual Basic, functions to access the graphics library of *STATISTICA* can be easily incorporated into *STATISTICA* Visual Basic programs via a hierarchically organized ***Function Browser***. It contains short descriptions of all functions and options to insert them directly into the source code of your program (i.e., into the *STATISTICA* Visual Basic Editor, see page 231).

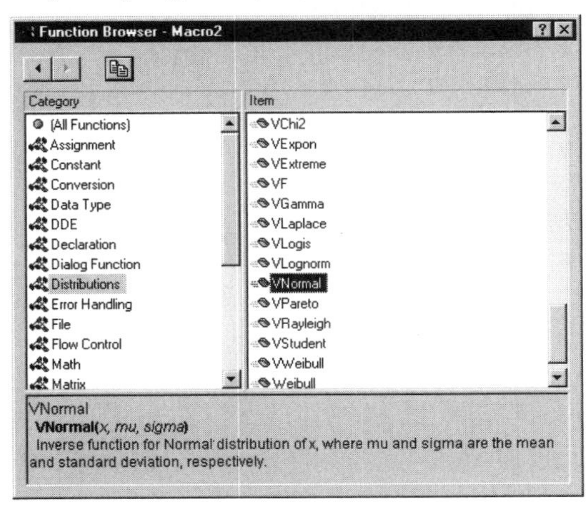

For more information on accessing the graphics libraries of *STATISTICA* via the *STATISTICA* Visual Basic programming language, refer to the *Electronic Manual*.

CUSTOMIZING *STATISTICA*

CHAPTER

Customizing *STATISTICA*

STATISTICA offers the flexibility of fully customizable user interfaces and supports the necessary adjustment of the standard user interface to better suit your specific needs. In fact, *STATISTICA* "anticipates" your needs in that it remembers various choices as you make them. For example, if you launch an analysis from the ***Advanced*** tab on an analysis specification dialog, the ***Advanced*** tab will be selected (instead of the ***Quick*** tab) the next time you display that dialog.

Practically all aspects of the user interface can be customized starting with such elementary controls as the menus, toolbars, and the keyboard. The process for customizing these screen components is quick and straightforward (for example, see the illustration of customizing the toolbar on page 143). You can set both global and local customizations for graphs, spreadsheets, workbooks, reports, etc., and maintain different configurations of *STATISTICA* (for a single user as well as for network users). You can also define entirely new user interfaces (see pages 143 and 144).

CUSTOMIZATION OF THE INTERACTIVE USER INTERFACE

As mentioned before, *STATISTICA* contains facilities to define entirely new user interfaces (see page 143), including the Internet browser-based user interfaces (see page 145). However, practically all aspects of the default, interactive user interface can also be adjusted easily in a variety of ways. For example, you can add to the default options, simplify them, or keep changing them as your needs change. Depending on the requirements of the tasks to be performed, as well as your personal preferences for particular "modes" of work (and aesthetic choices), you can suppress all icons, toolbars, status bars, long menus, workbook facilities, drag-and-drop facilities, dynamic (automatic) links between graphs and data, 3D effects in tables, and 3D effects in dialog boxes; request "bare-bones" sequential output with simple, paper-white spreadsheets and monochrome graphs; and set the system to automatically maintain no more than one simple report at a time (see the left panel in the illustration below);

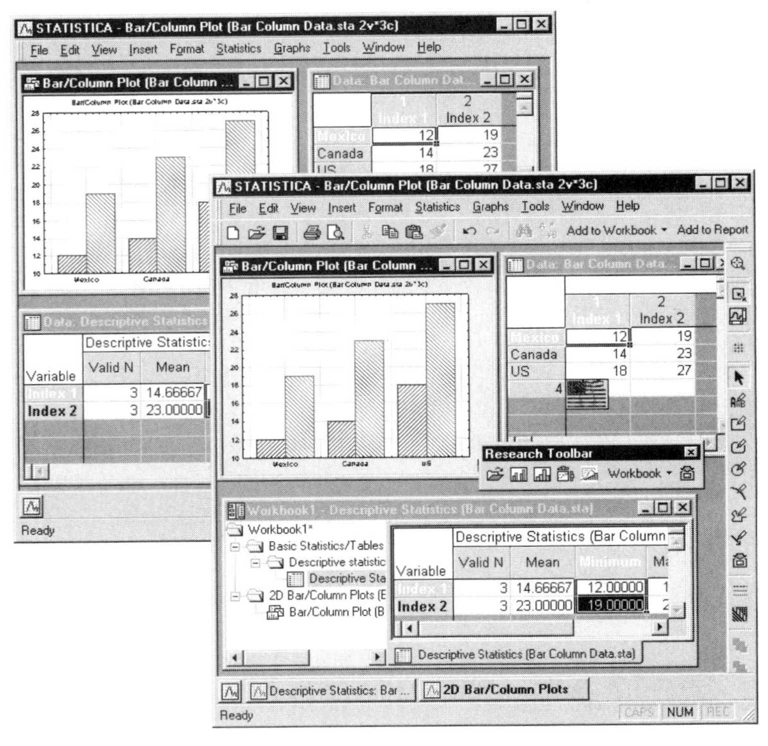

or alternatively, you could define elaborate local and global toolbars; take full advantage of all special tools and controls, icons, toolbars, macros (e.g., assign particular tasks to specific new menu options, toolbars, or keys), elaborate multimedia tables, workbook facilities, and drag-and-drop facilities; establish multiple dynamic (automatic) links between graphs and data and internal links between graphical objects; customize the output windows with colors, special fonts, and highlights; adjust the default graph styles and their display modes; and send the results to separate hierarchically organized workbooks to create an elaborate, "multi-layered" data analysis environment that facilitates the exploration of complex data files and allows you to compare different aspects of the output (see the right panel in the illustration, above).

CUSTOMIZATION OF DOCUMENTS

There is a variety of comprehensive, specialized tools to customize the layout and operation of *STATISTICA* documents (see Chapter 6 – *STATISTICA Documents*, page 175). For example, *STATISTICA* has a comprehensive system of managing defaults of every aspect of graphs and combining customizations into hierarchically organized "styles." Similarly, you can create custom layouts and formats for spreadsheets (multimedia tables) and even customize events (e.g., what happens when you double-click on a table). See the *Electronic Manual* for further details.

LOCAL VS. PERMANENT CUSTOMIZATIONS

Many aspects of the appearance of *STATISTICA* can be adjusted from both the ***View*** and ***Tools*** menus. Each of these two methods, however, has a different function.

View menu. The changes specified from the ***View*** menu affect the current appearance of *STATISTICA* (e.g., hides the toolbar) or the current document window (e.g., changes the font in the spreadsheet).

Tools menu. The options available via the ***Tools - Options*** menu (discussed in more detail in the next section) are used to adjust the permanent program defaults. Note, however, that the global options that are applicable to documents of a particular type (e.g., a graph or a spreadsheet) will not change the current document. Instead, they will only be stored as program defaults that will affect the creation of the next (i.e., new) document of the respective type.

For example, if you change the ***Default Spreadsheet Layout*** on the ***Spreadsheets*** tab of the ***Options*** dialog (accessible from the ***Tools - Options*** menu), you will see the new Spreadsheet Layout applied only when you create a new spreadsheet. However, these

defaults will not affect any files opened from the disk because those spreadsheets are displayed with the specific appearance with which they were previously saved (use the ***View*** menu to customize the existing objects).

GENERAL DEFAULTS

Customization of the general system defaults. The general default settings of *STATISTICA* can be adjusted using the options on the tabs of the ***Options*** dialog (accessible via the ***Tools - Options*** menu). They control:

- The general aspects of the behavior of *STATISTICA* (such as maximizing *STATISTICA* on startup, workbook and report facilities, file locations, custom lists, etc.),
- The way in which the output is produced (e.g., in workbooks, reports, etc.),
- The general appearance of the application window (icons, toolbars, etc.), and
- The appearance of document windows.

The ***General*** tab of the ***Options*** dialog is shown in the next illustration.

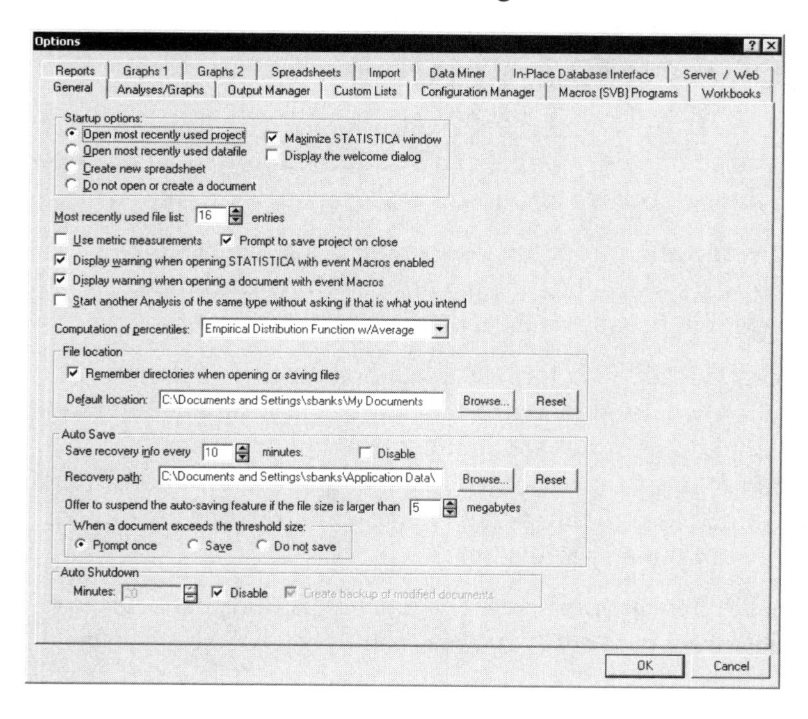

All these and other general settings are accessible regardless of the type of document that is currently active (e.g., a spreadsheet or a graph). For more information about a specific tab, see the *Electronic Manual*.

Switching between alternative sets of defaults (configurations). Options are provided on the ***Configuration Manager*** tab of the ***Options*** dialog that enable you to maintain "libraries" of settings and switch between them for different projects (or users). For further details, see the ***Configuration Manager*** tab on page 226 and in the *Electronic Manual*.

GRAPH CUSTOMIZATION

Interactive graph customization. The customization options in *STATISTICA* graphics include hundreds of features and tools that can be used to adjust every detail of the display and associated data processing. These options are arranged in a hierarchical manner, so those used most often are accessible directly via shortcuts by double-clicking or right-clicking on a specific element of the graph.

Permanent settings and automation options. The initial (default) settings of all graph features can be easily adjusted so that even the default appearance and behavior of *STATISTICA* graphs will match your specific needs and/or will require very little intervention on your part. Various aspects of *STATISTICA* Graphs can be permanently adjusted by using:

1. the ***Options*** dialog (accessible from the ***Tools - Options*** menu),

2. the comprehensive system of graph styles,

3. user-defined graphs, and

4. *STATISTICA* Visual Basic.

These facilities are briefly reviewed in Chapter 7 – *Graphs* (page 198). For more information, please refer to the *Electronic Manual*.

There are no limits to how "deeply customized" your *STATISTICA* custom graphs can be, because *STATISTICA* Visual Basic (with all its powerful custom drawing tools as well as the *STATISTICA*-based library of graphics procedures) can be used to produce virtually any graphics or multimedia output supported by the contemporary computer hardware. Those custom developed displays or multimedia

output can be assigned to *STATISTICA* toolbars, menus, or dialogs and become a permanent part of "your" *STATISTICA* application.

MAINTAINING DIFFERENT CONFIGURATIONS OF *STATISTICA*

STATISTICA stores all program settings when you exit the program, and restores them the next time you start the application. You can create different configurations of these settings by using the options on the ***Configuration Manager*** tab of the ***Options*** dialog (accessible via the ***Tools - Options*** menu). With the configuration manager, you can save the current program state into a new or existing configuration, or you can restart *STATISTICA* using a different configuration. Other options include the ability to import or export configurations to a separate file so they can be shared among *STATISTICA* installations.

CUSTOMIZED CONFIGURATIONS FOR INDIVIDUAL USERS ON A NETWORK

The same principle described in the previous paragraph applies to network installations of *STATISTICA*. On a network, *STATISTICA* is installed in only one location (on a server), but each user can still configure *STATISTICA* differently because the setting configuration information is stored locally. Note that you need to choose ***Network Installation*** in the ***STATISTICA Setup*** program in order to install it properly on a non-local drive (network server). Note that a network version of *STATISTICA* is necessary to assure its reliable operation when used by more than one user at a time or even one user if *STATISTICA* is not installed on the local system.

CHAPTER

STATISTICA VISUAL BASIC

STATISTICA Visual Basic

The industry standard *STATISTICA* Visual Basic (SVB) language (integrated into *STATISTICA*) provides another user interface to the functionality of *STATISTICA*, and it offers incomparably more than just a "supplementary application programming language" that can be used to write custom extensions.

SVB takes full advantage of the object model architecture of *STATISTICA* and is used to access programmatically every aspect and virtually every detail of the functionality of *STATISTICA*. Even the most complex analyses and graphs can be recorded into Visual Basic macros and later be run repeatedly or edited and used as building blocks of other applications. SVB adds an arsenal of more than 13,000 new functions to the standard comprehensive syntax of Visual Basic, thus comprising one of the largest and richest development environments available.

Applications for *STATISTICA* Visual Basic programs. *STATISTICA* Visual Basic programs can be used for a wide variety of applications, from simple macros recorded to automate a specific (repeatedly used) sequence of tasks, to elaborate custom analytic systems combining the power of optimized procedures of *STATISTICA* with custom developed extensions featuring their own user interface. When properly licensed, scripts for analyses developed this way can be integrated into larger computing environments or executed from within proprietary corporate software systems or Internet or intranet portals.

SVB programs can also be attached to virtually all important "events" in a *STATISTICA* analysis such as opening or closing files, clicking on cells in spreadsheets, etc.; in this manner, the basic user interface of *STATISTICA* can be highly customized for specific applications (e.g., for data entry operations, etc.).

RECORDING *STATISTICA* VISUAL BASIC (SVB) MACROS (PROGRAMS)

Analysis Macros, Master (Log) Macros, and Keyboard Macros

STATISTICA provides a comprehensive selection of facilities for recording macros, i.e., *STATISTICA* Visual Basic (SVB) programs, to automate repetitive work or to be used as a means to automatically generate programs for further editing and modification. The macro programs recorded by these facilities can be saved to be run "as is," or they can be used as the "building blocks" for more complex and highly customized Visual Basic application programs. Analysis Macros and Master Macros follow the identical syntax and can later be modified, but because of the different ways in which each of them is created, they offer distinctive advantages and disadvantages for specific applications.

Analysis Macros. Simple Analysis Macros automatically record the settings, selections, and chosen options for a specific analysis. (Note that the term "analysis" in *STATISTICA* denotes one task selected either from the ***Statistics***, ***Data Mining***, or ***Graphs*** menus, which can be very small and simple (e.g., one scatterplot requested from the ***Graphs*** menu), or very elaborate (e.g., a complex structural equation modeling analysis selected by choosing that command from the ***Statistics*** menu, and involving hundreds of output documents). After selecting any of the statistical commands from the ***Statistics*** or ***Data Mining*** menus or graphics commands from the ***Graphs*** menu, all actions such as variable selections, option settings, etc., are recorded "behind the scenes"; at any time you can transfer this recording (i.e., the Visual Basic code for that macro) to the Visual Basic Editor window. The ***Create Macro*** command is available from every analysis via the drop-down menu displayed by clicking the ***Options*** button or the shortcut menu (accessed by right-clicking the analysis button) when the analysis is minimized.

Master Macros (Logs). You can record a Master Macro or Master Log of an entire session, which can consist of one or many analyses. This recording will "connect" analyses performed with various analysis options from the , ***Data Mining***, or ***Graphs*** menus. However, unlike simple Analysis Macros, you can turn the recording of Master Macros on or off. The Master Macro recording will begin when

you turn on the recording [by selecting ***Start Recording Log of Analyses (Master Macro)*** from the ***Tools - Macro*** submenu], and it will end when you stop the recording (by selecting ***Stop Recording*** from the ***Tools - Macro*** submenu). In between these actions, all file selections and data management operations are recorded, as are the analyses and selections for the analyses, in the sequence in which they were chosen.

Keyboard Macros. This type of macro recording stores the sequences of keyboard input. When you select ***Start Recording Keyboard Macro*** from the ***Tools - Macro*** submenu, *STATISTICA* will record the actual keystrokes entered via the keyboard. When you stop recording, a *STATISTICA* Visual Basic editor window opens with a simple program containing a single **SendKeys** command with symbols that represent all the different keystrokes performed during the recording session. Note that this type of macro is very simple in the sense that it will not record any context in which the recorded keystrokes are pressed and will not record their meaning (i.e., commands these keystrokes trigger), but this feature makes them useful for specific applications, e.g., to automate entering text, such as titles, selection conditions, etc.

STATISTICA Visual Basic Editor and Debugger. Programs can be written from scratch using the *STATISTICA* Visual Basic professional development environment, which features a program editor with a powerful debugger (with breakpoints, etc.) and many facilities that aid in efficient code building. These facilities are described in detail in the *STATISTICA Electronic Manual*.

When editing macro programs by typing in Visual Basic commands or program commands specific to SVB, the editor displays type-ahead help to illustrate the appropriate syntax. Help on the members and functions for each class (object) is also provided in-line.

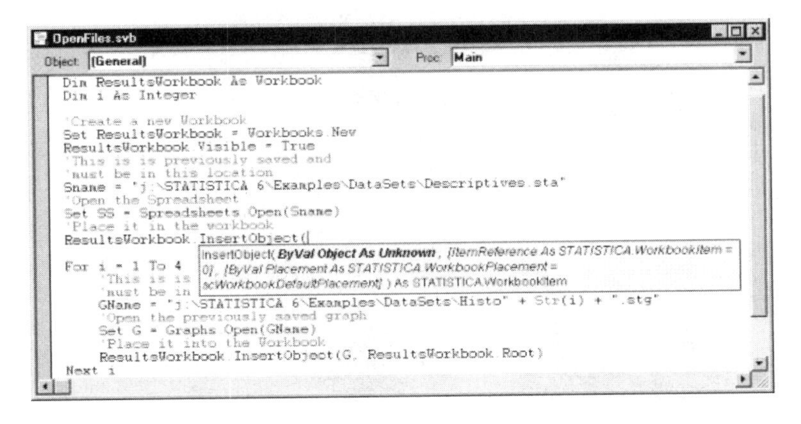

When executing a program, you can set breakpoints in the program, step through it line by line, and observe and change the values of variables in the macro program as it is running.

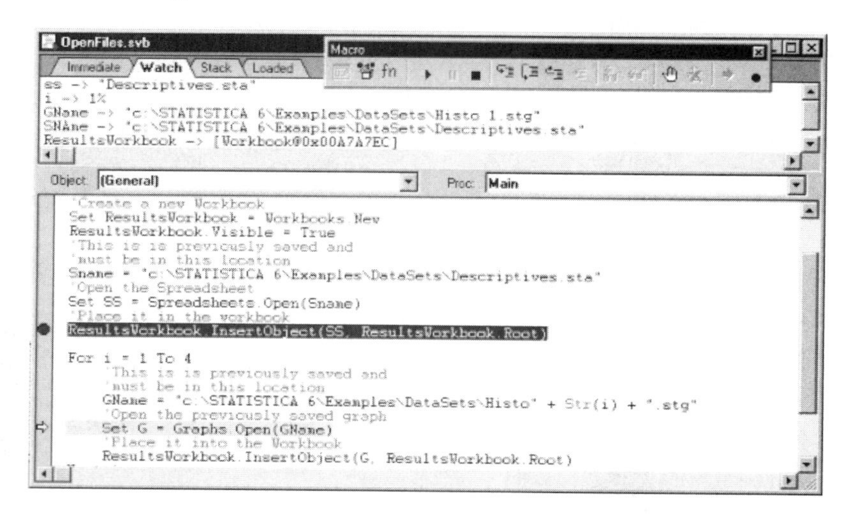

Also available is an interactive dialog editor that enables you to build dialog boxes.

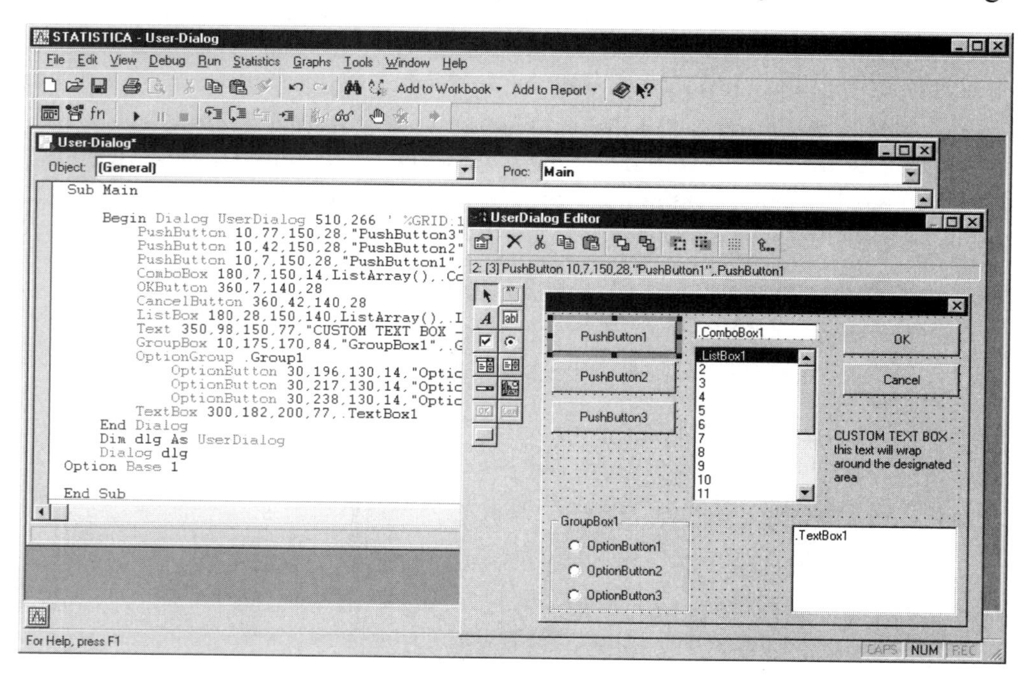

To summarize, *STATISTICA* Visual Basic is not only a powerful programming language, but it represents a very powerful, professional programming environment for developing simple macros as well as complex custom applications.

Visual Basic from other applications. SVB programs can also be developed by enhancing Visual Basic programs created in other applications (e.g., Excel) by calling *STATISTICA* functions and procedures.

Executing *STATISTICA* Visual Basic Programs

STATISTICA Visual Basic programs can be executed from within *STATISTICA,* but because of the industry standard compatibility of SVB, you can also execute its programs from any other compatible environment (e.g., Excel, Word, or a stand-alone Visual Basic language). In practice, you would typically call *STATISTICA* functions from Visual Basic in another application. Note, however, that when you run an SVB program or attempt to call *STATISTICA* functions from any other application, all calls to the *STATISTICA* specific functions (as opposed to the generic functions of MS Visual Basic) will be executed only if the respective *STATISTICA* libraries are present on the computer where the execution takes place. That is, you must be a licensed user of the respective *STATISTICA* libraries of procedures. Note that this large library of *STATISTICA* functions (more than 13,000 procedures) is transparently accessible not only to Visual Basic, but also to calls

from any other compatible programming language or environment, such as C/C++, C#, or Delphi.

Performance of *STATISTICA* Visual Basic programs. While the obvious advantages of Visual Basic (compared to other languages) are its ease of use and familiarity to a very large number of computer users, the possible drawback of Visual Basic programs is that they do not perform as fast as applications developed in lower-level programming languages (such as C). However, that potential problem does not apply to SVB applications, especially those that rely mostly on executing calls to *STATISTICA's* analytic, graphics, and data management procedures. These procedures fully employ *STATISTICA* technology and perform at a speed comparable to running the respective procedures in *STATISTICA* directly.

Structure of *STATISTICA* Visual Basic. *STATISTICA* Visual Basic consists of two major components: 1) The general Visual Basic programming environment with facilities and extensions for designing user interfaces (dialogs) and file handling, and 2) the *STATISTICA* libraries with thousands of functions that provide access to practically all functionality of *STATISTICA*.

The Visual Basic programming environment follows the industry standard syntax conventions of the Microsoft Visual Basic Language; the few differences pertain mostly to the manner in which dialogs are created (see *Custom Dialogs* and *Custom User Interfaces* in the *STATISTICA Electronic Manual*), and are designed to offer programmers/developers more flexibility in the way user interfaces are handled in complex programs. In the SVB programming environment, dialogs can be entirely handled inside separate subroutines, which can be flexibly combined into larger multiple-dialog programs; MS Visual Basic is form based, where the forms or dialogs, and all events that occur in the dialogs, are handled in separate program units.

Attaching Macros to Toolbars and Menus

A *STATISTICA* Visual Basic program can be saved and then attached to a custom toolbar or menu item. This enables you to easily customize and extend the operation and appearance of *STATISTICA* with your own custom macros. To utilize these facilities, save the macro by selecting ***Save As Global Macro*** from the ***File*** menu. Then, to customize the menus and/or tool bars, select ***Customize*** from the ***Tools*** menu to display the ***Customize*** dialog. To add the macro to a menu or toolbar, choose the ***Command/Macros*** tab, and select ***Macros*** from the ***Categories*** list. All

your Global Macros will be listed in the ***Commands*** section of the tab. In the following illustration, macro *Test1* has been saved as a Global Macro.

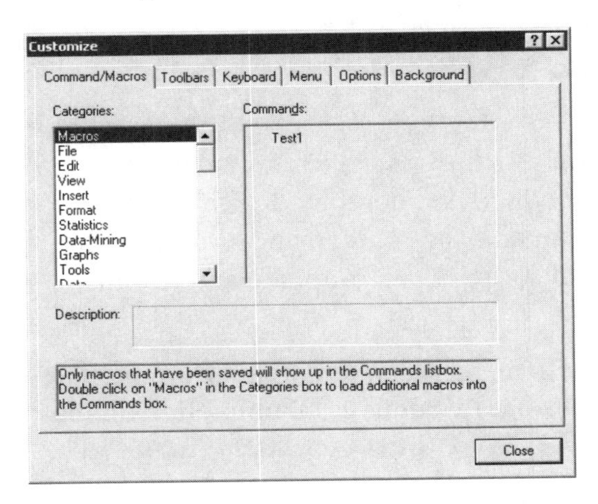

You can then select and drag the specific item from the ***Commands*** list onto any menu or toolbar. Note that as your mouse pointer hovers over a menu, the menu will expand, enabling you to insert the item in any submenu as well. Once the macro is placed on the menu or toolbar while the ***Customize*** dialog is displayed, you can right-click the macro and change the appearance and text of the item, as well as add icons.

Running Macros from a command line. With *STATISTICA*, you can execute SVB programs from the command line by using the /RunMacro= command line parameter. The syntax is:

```
statist.exe /RunMacro=macroname
```

where “macroname” is the file name of the macro. If a full path is not specified, *STATISTICA* will attempt to run the macro from the application’s currently selected directory (which is Windows default behavior).

If the macro does not make the application or any document visible (through the `Application.Visible = True`, or similar document properties), then the *STATISITCA* instance will automatically shut down when complete. If the application is made visible, then the application will remain visible after the macro completes, and you will need to shut down the program

EXAMPLE: RECORDING AN ANALYSIS

This example illustrates how to record an analysis into a script that can be executed to re-run the analysis. Then the script will be edited and combined with another script to create a customized script that can run analyses on demand. Additionally, this example shows how you can use attached scripts to auto-update and re-run analyses from results workbooks.

Start by opening the example *Adstudy* data set. Select ***Open Examples*** from the ***File*** menu item to display the ***Open a STATISTICA Data File*** dialog. Double-click on the *Datasets* file, and then select and open the *STATISTICA* data set *Adstudy.sta*.

Then, select ***Basic Statistics/Tables*** from the ***Statistics*** menu. In the ***Basic Statistics and Tables*** dialog, select ***Descriptive statistics***.

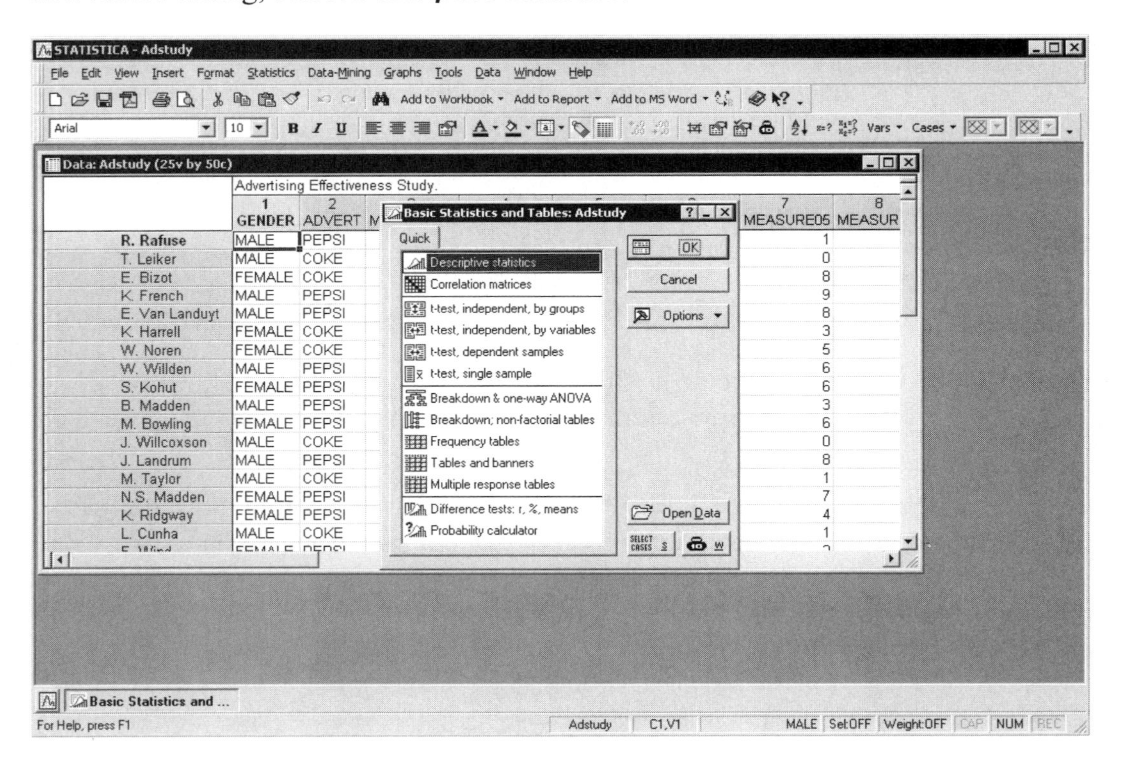

Click the ***OK*** button, and the ***Descriptive Statistics*** dialog will be displayed.

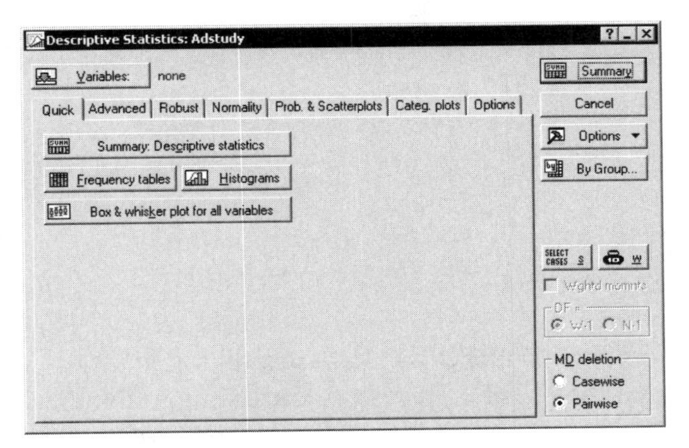

Click the ***Variables*** button to display the ***Select the variables for the analysis*** dialog. Select variables *MEASURE01* through *MEASURE23* by clicking *MEASURE01* and dragging to *MEASURE23*, and then click ***OK***. In the ***Descriptive Statistics*** dialog, click the ***Advanced*** tab, and note the numerous options available.

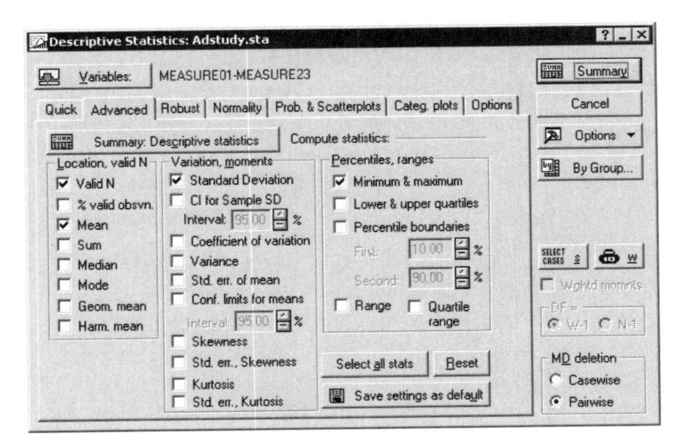

For this example, we will leave all options at their default. Click the ***Summary*** button to display the descriptive statistics for the selected variables.

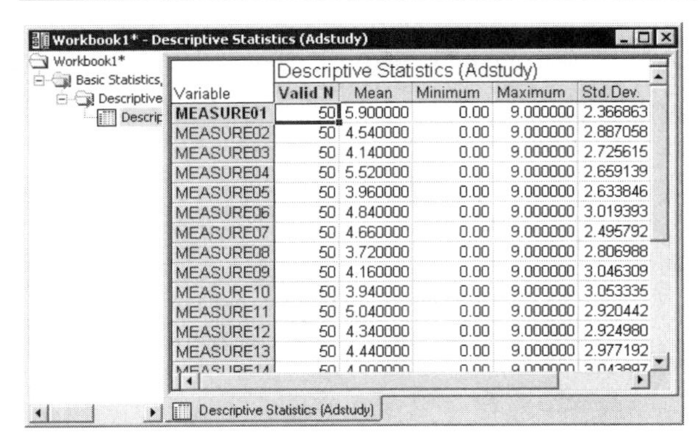

Variable	Valid N	Mean	Minimum	Maximum	Std.Dev.
MEASURE01	50	5.900000	0.00	9.000000	2.366863
MEASURE02	50	4.540000	0.00	9.000000	2.887058
MEASURE03	50	4.140000	0.00	9.000000	2.725615
MEASURE04	50	5.520000	0.00	9.000000	2.659139
MEASURE05	50	3.960000	0.00	9.000000	2.633846
MEASURE06	50	4.840000	0.00	9.000000	3.019393
MEASURE07	50	4.660000	0.00	9.000000	2.495792
MEASURE08	50	3.720000	0.00	9.000000	2.806988
MEASURE09	50	4.160000	0.00	9.000000	3.046309
MEASURE10	50	3.940000	0.00	9.000000	3.053335
MEASURE11	50	5.040000	0.00	9.000000	2.920442
MEASURE12	50	4.340000	0.00	9.000000	2.924980
MEASURE13	50	4.440000	0.00	9.000000	2.977192

When you produce the results workbook, the ***Descriptive Statistics*** dialog is automatically minimized so you can see the results. To restore the dialog, click the ***Descriptive Statistics*** button on the Analysis Bar in the lower-left of the screen.

While you are running this analysis, *STATISTICA* automatically records all the analysis steps behind the scenes. You can now produce a *STATISTICA* Visual Basic (SVB) macro to re-create this analysis. In the ***Descriptive Statistics*** dialog, click the Options button, and select ***Create Macro*** from the drop-down menu. The ***New Macro*** dialog will be displayed, where you can name the macro and enter a description. Leave all the entries at their defaults, and click ***OK***. An SVB macro window will be displayed, containing the recorded *Descriptive Statistics* session.

To run this macro, select ***Run Macro*** from the ***Run*** menu or press F5 on your keyboard. The exact descriptive statistics results that were generated in the initial analysis will be reproduced.

Look at the SVB macro for a moment. Toward the top, one of the lines is:

```
Set newanalysis = Analysis (scBasicStatistics, ActiveDataSet)
```

This is telling the macro that it is going to run the *Basic Statistics* analysis, and that it will be using the "active" data set, that is, the spreadsheet that is currently selected when the macro runs.

Down a few lines is a section that starts with:

```
With newanalysis.Dialog
```

and under that are properties such as:

```
.Mean = True
```

These properties correspond to all the options that were available on the different tabs of the ***Descriptive Statistics*** dialog. Every option on the dialog is represented by a property, and all the current settings are recorded. If you decide to include a ***Median*** and the ***Sum*** of each of the variables, it is easy to add this to the SVB macro; just find the lines that read:

```
.Median = False
```

and

```
.Sum = False
```

and change these to:

```
.Median = True
```

and

```
.Sum = True
```

Now, run the macro again by pressing F5. A new results spreadsheet will be added to the workbook, this time with new columns of *Median* and *Sum*:

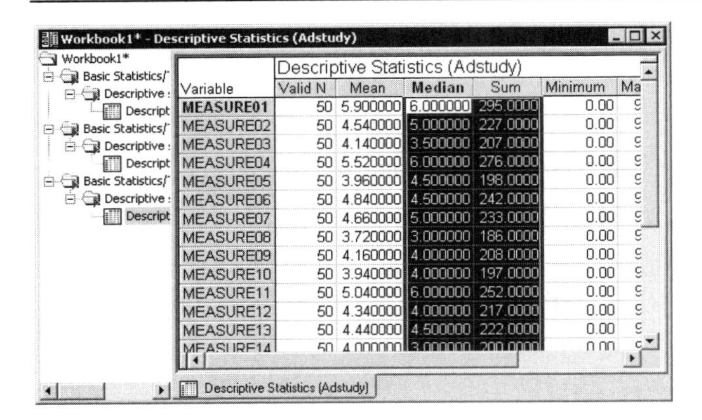

Let's keep the macro window open and start a new analysis on the same sample data set. Select the *Adstudy* spreadsheet to bring it to the front. From the ***Graphs - 2D Graphs*** submenu, select ***Normal Probability Plots*** to display the ***Normal Probability Plots*** dialog.

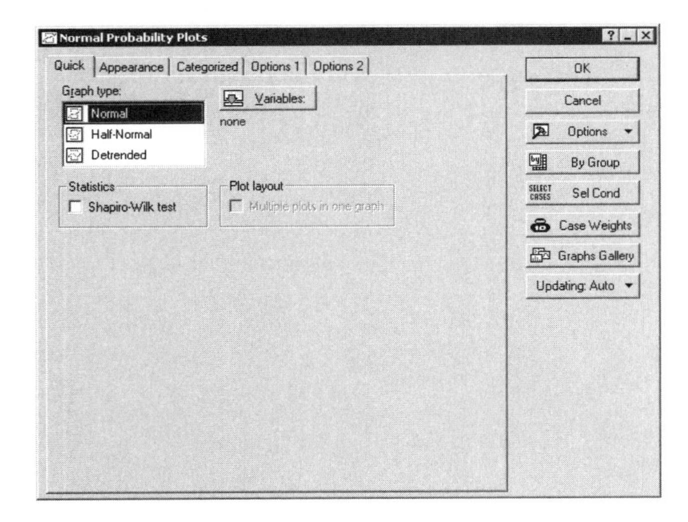

Click the ***Variables*** button, and on the ***Select Variables for Probability Plot*** dialog, select variables *MEASURE01* through *MEASURE03*. Click ***OK*** to close this dialog, and click ***OK*** in the ***Normal Probability Plots*** dialog. Three *Probability Plot* graphs will be placed in the results workbook, one for each of the three variables that were selected.

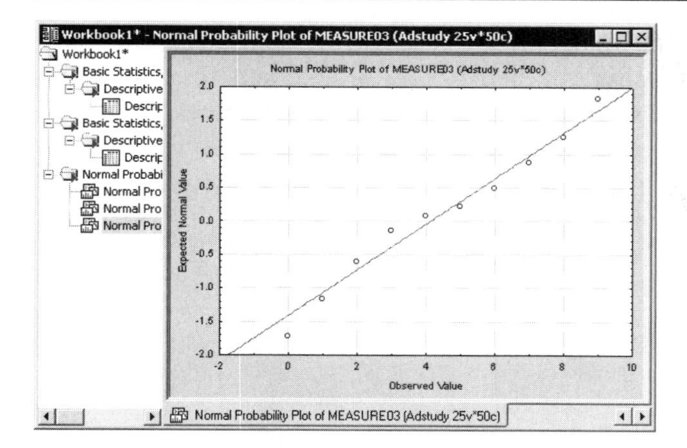

The steps of the *Probability Plot* analysis were recorded just as they were for the *Descriptive Statistics* analysis. To create a new macro with these steps, bring the ***Normal Probability Plot*** dialog to the front by clicking that button on the Analysis Bar in the lower-left of the screen, click the Options button, and select ***Create Macro*** from the drop-down menu. In the ***New Macro*** dialog, click ***OK***, and a new SVB Macro window is opened with the recorded *Probability Plot* script.

As with the *Descriptive Statistics* analysis, all the options selected in the ***Probability Plot*** dialog are specified as properties within the macro. For instance, to change this from a *Normal Probability Plot* to a *Half Normal Probability Plot*, locate the following line:

```
.GraphType = scProbNormal
```

and change it to:

```
.GraphType = scProbHalfNormal
```

Also, let's expand the variables to include variable *MEASURE04*. To do this, find the following line:

```
.Variables = "3-5"
```

This line corresponds to the variables selected for the plots. Since we selected *MEASURE01* through *MEASURE03*, and these are variable numbers *3* through *5* from the data set, this string was recorded. To add *MESURE04* (variable number *6*), change this line to:

```
.Variables = 3-6
```

Now run the macro by pressing F5. Four new graphs are produced as *Half Normal Probability Plots* for variables *MEASURE01* through *MEASURE04*.

This example has demonstrated how you can run any analysis, and then create a macro of the analysis that can be edited and played back. Additionally, this example has shown how these macros can be combined to make macros that are more complex. This is the building block of creating your own powerful customized analyses using the SVB language.

Rerunning Analyses from Results Workbooks

In the previous example, you learned that all analyses in *STATISTICA* will record the steps used to produce them, and these can be loaded into a macro that you can edit and run. When an analysis produces results that are placed in a workbook, *STATISTICA* automatically associates the recorded script's steps to the workbook folder that contains the results. This enables you to be able to either re-run the analysis or to resume an analysis.

Thus far, we have produced several instances of running both *Descriptive Statistics* and *Probability Plots*. The results workbook looks similar to the following illustration.

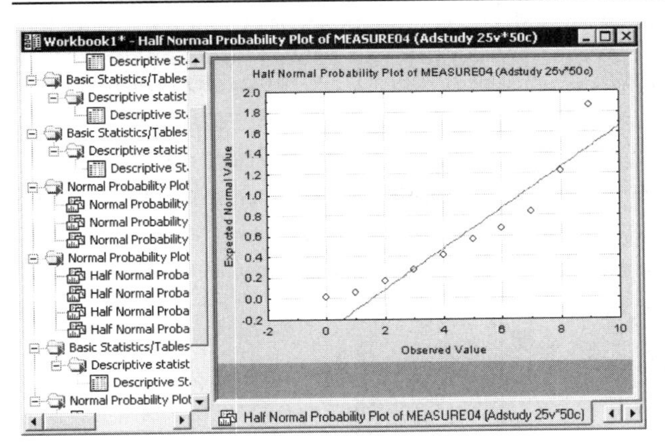

Notice that there is a red arrow on each workbook folder. This is an indicator that the script that produced the results in that folder has been attached to the folder. This enables *STATISTICA* to re-run or resume the analysis.

To re-run an analysis, right-click on one of the folders labeled *Descriptive statistics dialog*, and from the shortcut menu, select ***Re-run Analysis***. The ***Re-run Analysis*** dialog will be displayed.

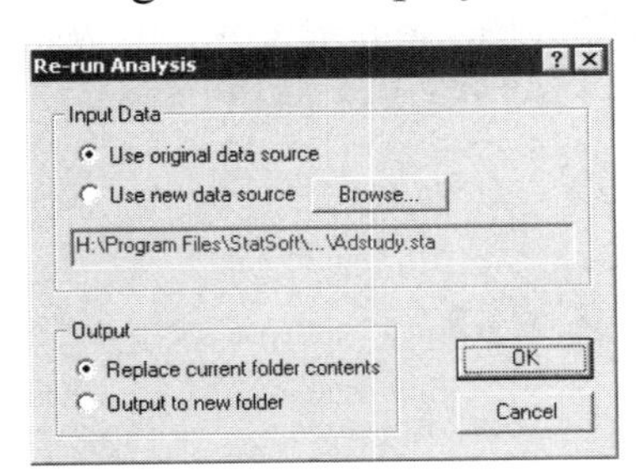

Here you can choose to ***Use original data source*** or ***Use new data source***. The latter option gives you the powerful ability to create "templates" that can then be applied to new data sources. In addition to specifying the data source, you can choose to ***Replace current folder contents*** or ***Output to new folder***. In this example, leave the defaults, and click ***OK***. You will see that the contents of the folder are briefly deleted and then added again as the analysis is re-run.

One purpose for this feature is the ability to update/re-run results from complex analyses if new data is entered into the spreadsheet. For instance, if the data in the open data file *Adstudy.sta* has been changed and the analysis is re-run, the new results will be calculated with the new data.

The resume analysis functionality enables you to bring an analysis back to the point before the results were generated, allowing you to select different options or continue an analysis in progress. Right-click the same *Descriptive statistics dialog* folder, and from the shortcut menu, select ***Resume Analysis***. The ***Resume Analysis*** dialog will be displayed. This dialog also contains options to specify the input data source (original or new). The ***Output*** options for the new results are to ***Output to current folder*** (as if this is just an extension of the previous analysis) or ***Output to new folder*** (as if this is a brand new analysis).

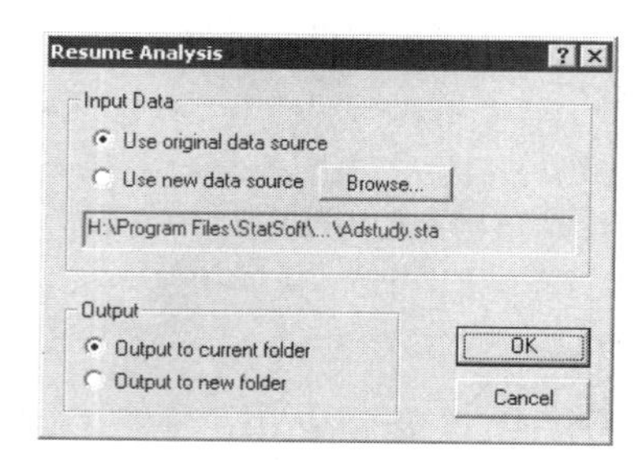

Leave the defaults as they are, and click ***OK***. The ***Descriptive Statistics*** dialog will be displayed, with all the options set to what was used when the selected output was created. Since the default was to ***Output to current folder***, clicking the ***Summary*** button will generate new output to the same folder.

ACTIVEX OBJECTS AND DOCUMENTS (A TECHNICAL NOTE)

The term ActiveX is used in different contexts, and its definitions stress different aspects of that concept. Its use within *STATISTICA*, however, can be grouped into two general categories: ActiveX objects and ActiveX documents.

ActiveX Objects. An ActiveX object is what was once referred to as an OLE (Object Linking and Embedding) object. At its heart is the Microsoft COM (Component Object Model) technology that allows objects to be accessed in a uniform manner. Through the use of standard protocols, objects created in one application can be stored and edited in a different application. To support this functionality, the containing object needs to be an ActiveX object client, and the application that initially created the object needs to be an ActiveX object server.

STATISTICA is both. As an ActiveX object client, *STATISTICA* allows you to embed and link objects from other applications in the spreadsheet, graph, and report windows. As an ActiveX object server, it allows you to embed and link spreadsheets and graphs into other applications.

ActiveX documents. ActiveX documents take the ActiveX controls one step further, in that they allow entire documents to be embedded into other applications. An ActiveX document container allows other application documents to be used within it, and an ActiveX document server allows its documents to be used within any ActiveX document container. Again, *STATISTICA* does both. *STATISTICA* Workbooks are ActiveX document containers, and allow documents from other ActiveX servers to be displayed within the workbook. Examples of this are Word and Excel; these documents can be used directly from within a *STATISTICA* Workbook. Similarly, *STATISTICA* Spreadsheets, Graphs, and Reports are ActiveX document servers, and they also can be placed within any ActiveX document container such as Microsoft Internet Explorer and Microsoft Binder.

Office Integration and ActiveX documents. The ActiveX document technology has special application with Word and Excel documents. *STATISTICA* can open these particular documents natively in their own windows within the *STATISTICA* workspace. This "Office Integration" enables you to use Excel documents as data sources, and Word documents as reports for analyses. When the documents are open in the *STATISTICA* Window, the appropriate menus and toolbars for Excel/Word are available for use.

OLE SUPPORT

STATISTICA supports the Object Linking and Embedding (OLE) conventions that are used to link values, text, graphs, or other objects in *STATISTICA* documents (e.g., spreadsheets, reports) to documents in other (Windows) applications. Technically speaking, you can establish OLE links between a "source" (or server) file (e.g., a Word document) and a *STATISTICA* document (the "client" file), so that when changes are made to the data in the source file, the data are automatically updated in the respective part of the *STATISTICA* document (client file). Additionally, *STATISTICA* can serve as a "source" (or server) file for other OLE compatible documents. In this way, you can link the values in a *STATISTICA*

Spreadsheet to a table in a Word document, so that the Word document updates when data in the spreadsheet are changed.

DDE SUPPORT

STATISTICA also supports the Dynamic Data Exchange (DDE) conventions. You can establish DDE links between a "source" (or server) file (e.g., an Excel spreadsheet) and a *STATISTICA* data file (the "client" file), so that when changes are made to the data in the source file, the data are automatically updated in the respective part of the *STATISTICA* Spreadsheet (client file). Thus, you can dynamically link a range of data in its spreadsheet to a subset of data in other (Windows) applications.

A common application for dynamically linking two files would be in industrial settings, where the *STATISTICA* data file is dynamically linked with a measurement device connected to the serial port (e.g., in order to automatically update specific measurements hourly). Like OLE, the procedure is in fact much simpler than it might appear and can be easily employed without technical knowledge about the mechanics of DDE.

STATISTICA QUERY

STATISTICA Query

Note: For an explanation of all technical terms used in this overview (e.g., ODBC, SQL, etc.), please refer to the glossary in the *STATISTICA Electronic Manual*, accessible by selecting ***STATISTICA Help*** from the ***Help*** menu.

OVERVIEW

STATISTICA Query is used to access data easily from a wide variety of databases (including many large system databases such as Oracle, MS SQL Server, Sybase, etc.) using Microsoft's OLE DB conventions. OLE DB is a powerful database technology that provides universal data integration over an enterprise's network, from mainframe to desktop, regardless of the data type. OLE DB offers a more generalized and more efficient strategy for data access than the older ODBC conventions because it allows access to more types of data and is based on the Component Object Model (COM).

STATISTICA Query supports multiple database tables; specific records (rows of tables) can be selected by entering SQL statements, which *STATISTICA* Query automatically builds for you as you select the components of the query via a simple graphical interface and/or intuitive menu options and dialogs. Therefore, an extensive knowledge of SQL is not necessary in order for you to create advanced and powerful queries of data in a quick and straightforward manner. Multiple queries based on one or many different databases can also be created to return data to an individual spreadsheet, and you can maintain connections to multiple external databases simultaneously.

QUICK, STEP-BY-STEP INSTRUCTIONS

The steps necessary to retrieve external data via *STATISTICA* Query are outlined below:

1. Select ***Create Query*** from the ***Data - Get External Data*** submenu (or from the ***File - Get External Data*** submenu) to display the ***Database Connection*** dialog. In this dialog, select a predefined database connection (the provider, data source location, and advanced settings of the server or directory on which the data resides). Note that if you have not already created the database connection, you can do so by clicking the ***New*** button on the ***Database Connection*** dialog. The ***Data Link Properties*** dialog will be displayed, which will guide you through a step-by-step wizard to create a database connection. For specific documentation when you are using the ***Data Link Properties*** dialog, press the F1 key on your keyboard to display the Microsoft Data Link Help®.

2. After you have selected a database connection and clicked the ***OK*** button on the ***Data Link Properties*** dialog, you will have access to *STATISTICA* Query in which you can create your SQL statement by specifying the desired tables, fields, joins, criteria, etc. (via the ***Table***, ***Join,*** and ***Criteria*** menus) to be included in your query.

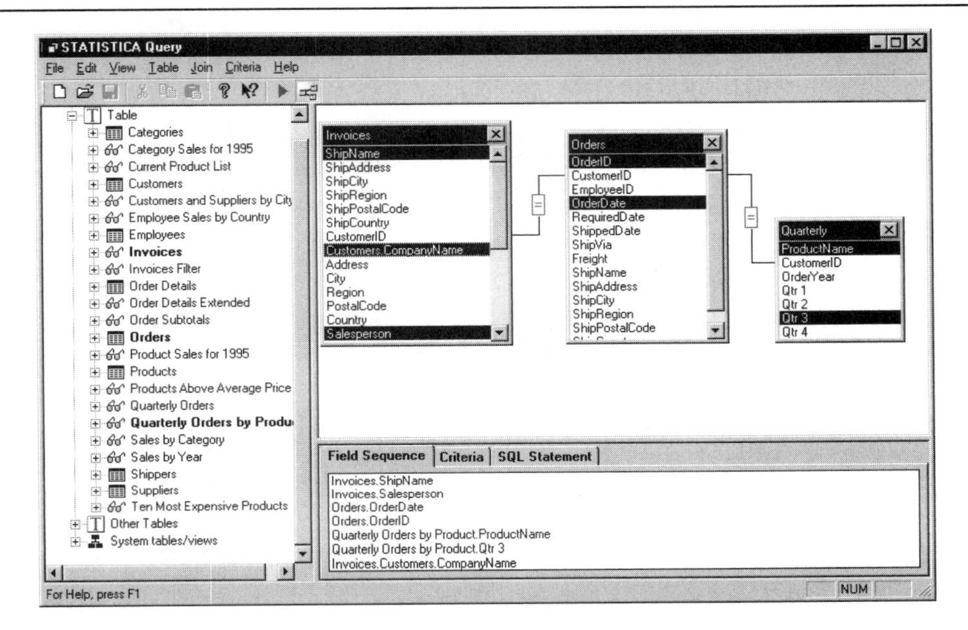

3. Once you have specified a query, select ***Return Data to STATISTICA*** from the ***File*** menu. The ***Returning External Data to Spreadsheet*** dialog will be displayed in which you can specify the name of the query, where you want *STATISTICA* Query to put the data that the query returns, and additional options.

See the *Electronic Manual* for further details.

IN-PLACE PROCESSING OF DATA ON REMOTE SERVERS (THE IDP TECHNOLOGY OPTION)

The query facilities (described in the previous sections), when offered as part of the enterprise versions of *STATISTICA* (see *STATISTICA* , page 284), are additionally enhanced by options to process data from remote servers "in-place," that is, without having to import them and create a local data file. This *In-Place Database Processing* (*IDP*) technology is particularly useful for processing extremely large data files where it can produce significant performance gains and enable *STATISTICA* users to process data files that exceed the storage capacity of the local device or even the *STATISTICA Server*.

Technical Note. The *IDP* technology is based on distributing processing architecture, where the queries are performed on the server side (using the server CPU resources) and the respective records sent to the *STATISTICA* computer where they are simultaneously (asynchronously) processed as they become available.

11

PROGRAMMING *STATISTICA* FROM .NET

CHAPTER

PROGRAMMING *STATISTICA* FROM .NET

Virtually every aspect of *STATISTICA* is exposed as a set of COM interfaces that are registered on a machine when *STATISTICA* is installed. Since .NET-based languages cannot communicate with COM directly, a wrapper class called the COM Interop can be utilized to integrate the *STATISTICA* libraries into your .NET project. The COM Interop layer is created automatically by the Visual Studio .NET IDE when you import a COM interface. The COM Interop layer handles all of the details regarding interacting with the COM libraries in .NET. With the COM Interop layer in place, the *STATISTICA* COM interfaces behave like any other .NET object.

ADDING THE *STATISTICA* OBJECT LIBRARY INTO YOUR .NET PROJECT

The .NET Interop layer is created automatically by adding the desired *STATISTICA* COM interfaces into your .NET project. *STATISTICA Object Library* is the base *STATISTICA* COM library. To add the *STATISTICA* Object Library to a .NET project, first select the desired .NET project in *Solution Explorer*, and then select ***Add References*** from the shortcut menu (accessed by right-clicking on the .NET project). The ***Add Reference*** dialog will be displayed.

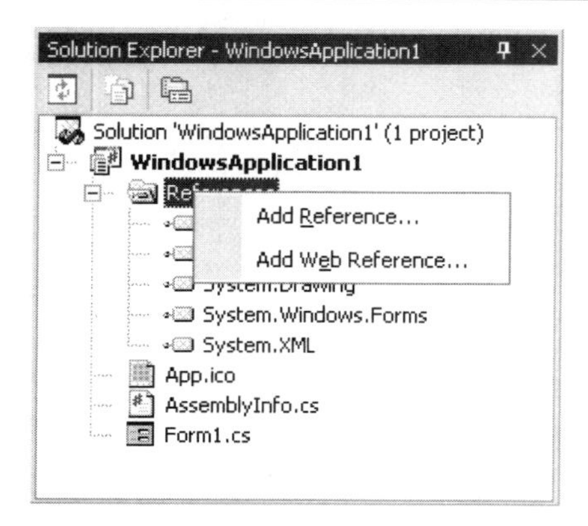

In the ***Add Reference*** dialog, select the ***COM*** tab. From the **Component Name** list, select *STATISTICA Object Library,* and click ***OK***.

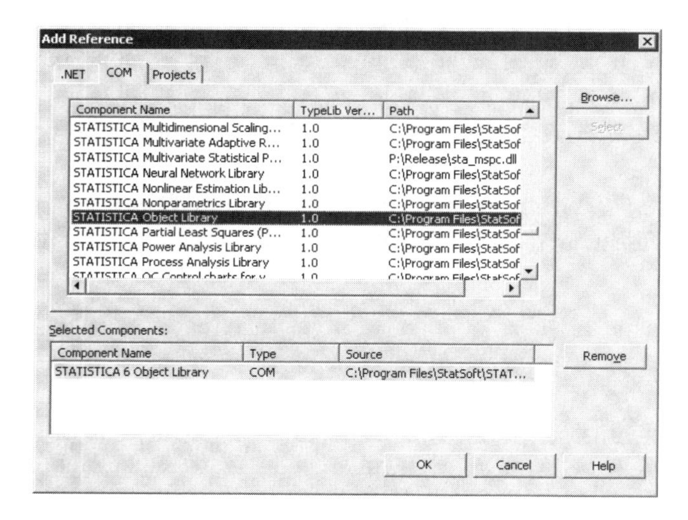

At this point, the necessary COM Interop library is created automatically. Under the project *References* node, you will now see the entry *STATISTICA*.

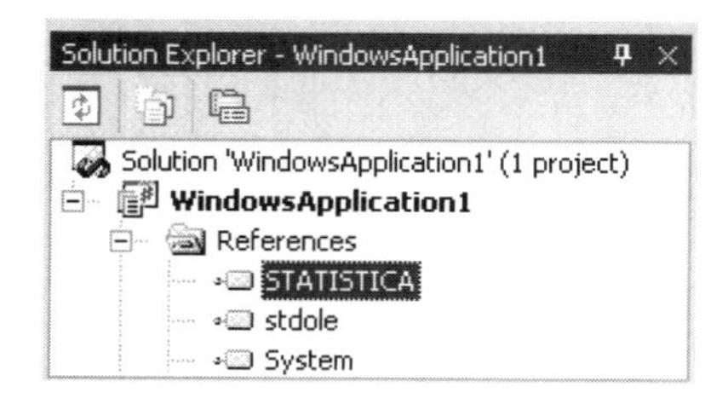

The file Interop.*STATISTICA.dll* is also added to the project output directory. The *STATISTICA* COM Interop library is stored in this file. To view the *STATISTICA* object library from your .NET project, right-click on the *STATISTICA* reference, and from the shortcut menu, select ***View in Object Browser***.

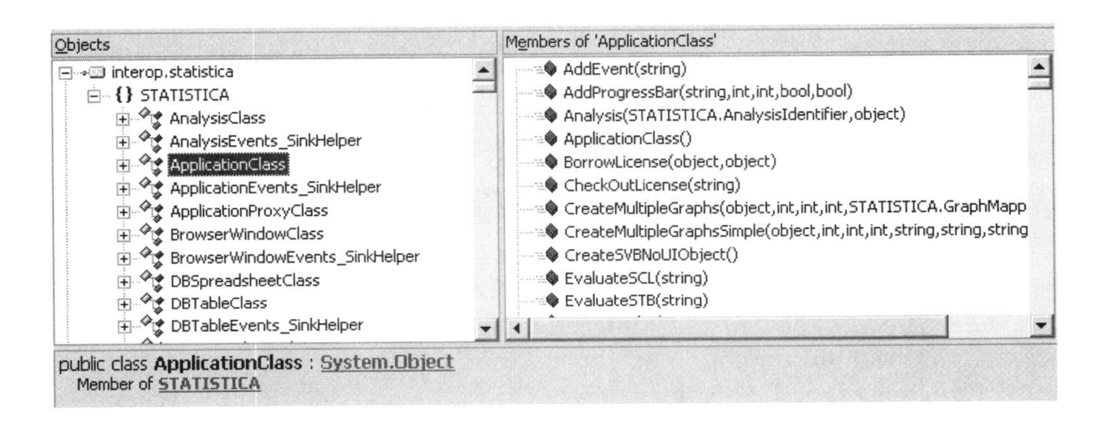

MANUALLY CREATING THE COM INTEROP LIBRARY

It is also possible to create the COM Interop library manually and import it into your .NET project. This gives you the ability to specify a different name for the Interop DLL as well as define a custom namespace. The program that enables you to create an Interop is *TLBIMP.EXE*. From a Visual Studio command prompt, execute *TLBIMP* with an initial parameter of the type library source. In the example below, the output DLL name and namespace are also specified.

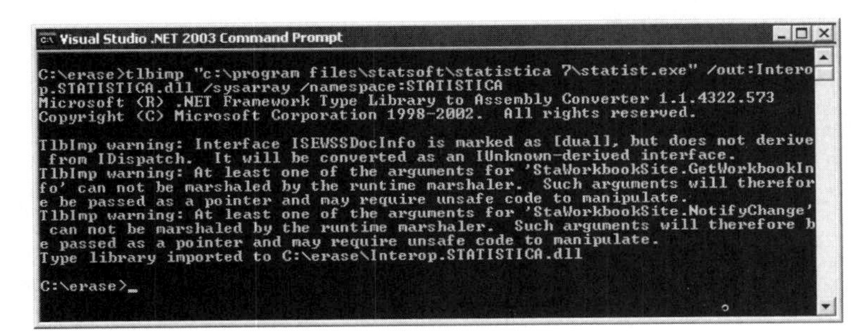

In this example, we reference the file *STATIST.EXE* since that executable contains the *STATISTICA Object Library* type library. Once the Interop DLL is generated, you can add it to your .NET project by selecting ***Add Reference*** from the *Solution Explorer* as before, but this time click the ***Browse*** button to select the newly created Interop DLL.

SUPPORTING MULTIPLE VERSIONS OF *STATISTICA*

To support multiple version numbers of *STATISTICA*, it is necessary to maintain separate *STATISTICA Object Library* Interop DLLs for each version number of *STATISTICA* you want to support. You can use the *TLBIMP* command to generate Interop DLLs against specific versions of *STATIST.EXE* and other DLLs. When distributing the application, make sure the correct version of the *STATISTICA* Interop DLL is deployed with your .NET application.

INSTANTIATING *STATISTICA*

Because of its COM architecture, *STATISTICA* can be incorporated into many different development environments. When using *STATISTICA* from an external development environment, it is necessary to have a top-level object called the application object. The application object is the application itself and will contain other objects (for example, spreadsheets and graphs), but access to these other objects is restricted unless the application object is running.

Assuming you are using the default namespace *STATISTICA*, the interface you should declare your variable as is *STATISTICA.Application*. To create an instance of *STATISTICA*, set your variable equal to `new STATISTICA.ApplicationClass()`.

```
STATISTICA.Application pApp = (STATISTICA.Application)
   new STATISTICA.ApplicationClass();

pApp.Visible = true;
```

When an instance of the *STATISTICA.ApplicationClass* is created, a *STATIST.EXE* process will be launched. This is equivalent to launching *STATISTICA* from the ***Start*** menu. The *STATISTICA* instance is initially hidden but can be made visible. Since it is a separate process, all calls to this instance are made out of process.

THE LIBRARY VERSION OF *STATISTICA*

In addition to the *STATISTICA.Application* object, there is also a lighter-weight, higher-performance version of the object called *STATISTICA.Library*. The *Library* version is licensed separately and therefore may not be available with your installation. It contains identical interfaces as the *STATISTICA.Application* library. Any existing code that uses the *Application* object can be replaced with the *Library* object.

The main restriction is that the *STATISTICA* user interface features are not available from the *Library* version. Therefore, in the example above, if the *Application* object was instantiated as a new *STATISTICA.LibraryClass*, it would not be possible to make the object visible (and show the *STATISTICA* interface).

The *Library* version of *STATISTICA* is loaded in-process, which means accessing its COM interfaces is more efficient than using the *Application* version of the object (which is loaded out of process). Since it is loaded in-process, multiple versions of the library cannot be instantiated. Normally, you would only instantiate one *Library* object or one *Application* object in your program.

APPENDIX

GETTING MORE HELP

GETTING MORE HELP

Electronic Manual

The most convenient place to get help and access a vast repository of information about *STATISTICA* is the *Electronic Manual*, which contains more than 100 Megabytes of references, illustrations, and examples.

This hypertext document offers much more than just an explanation of the options in *STATISTICA*. It includes countless examples, overviews, and illustrations with thousands of tips on how to optimize your work.

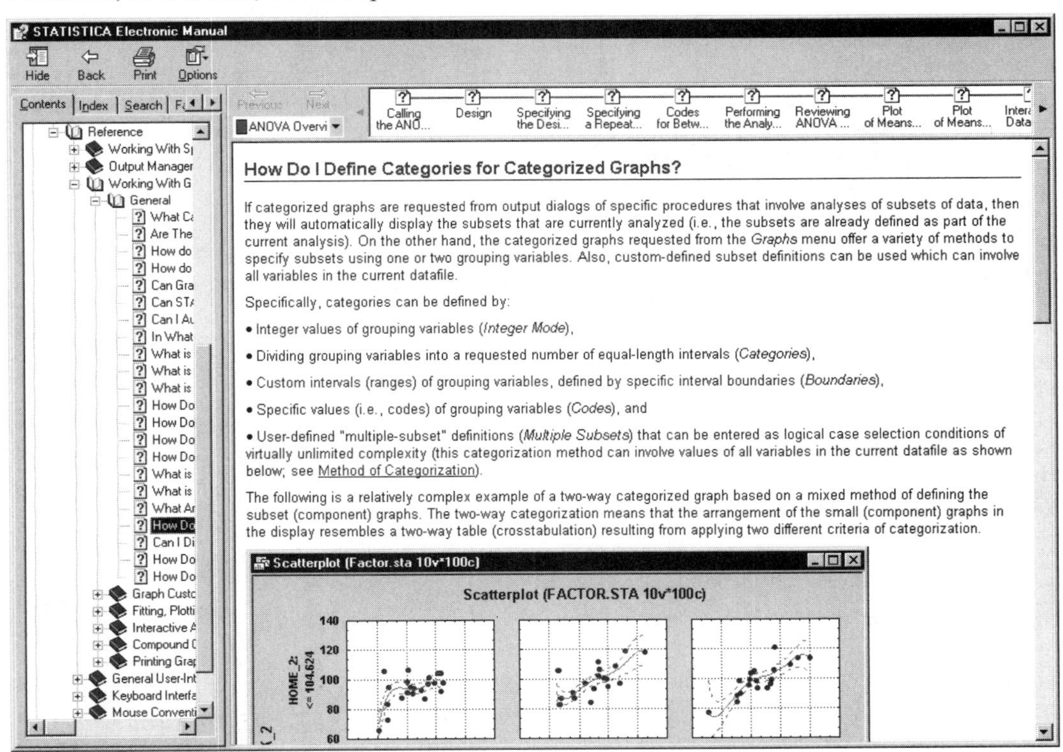

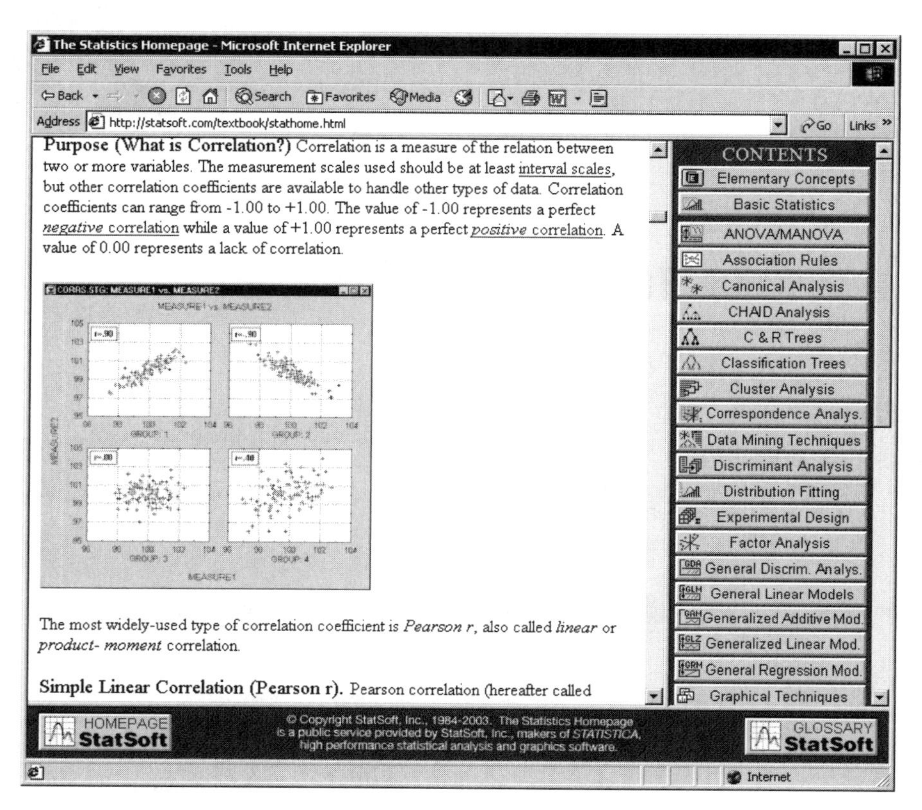

The *STATISTICA Electronic Manual* is extremely comprehensive. It offers a built-in ***Statistical Advisor*** (see page 33) supplemented with the complete contents of StatSoft's award-winning *Electronic Statistics Textbook* and *Glossary*.

StatSoft's *Electronic Statistics Textbook* has been recommended by *Encyclopedia Britannica* for its "Quality, Accuracy, Presentation, and Usability."

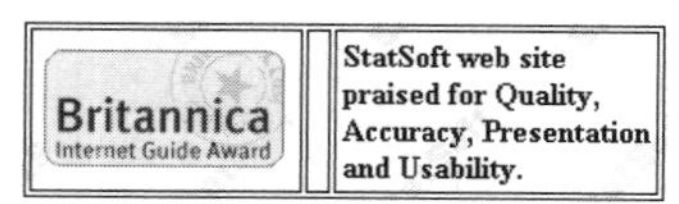

This unique textbook has been used for many years in educational and research activities at universities and research organizations worldwide.

Other Technical Support Resources and Facilities

Web site resources. StatSoft's Web site, one of the most visited Internet addresses related to data analysis, offers not only access to many resources that are useful for data analysis professionals in general, but it also includes:

- A continuously updated *Frequently Asked Questions* section, and
- A download area where users of the current version of *STATISTICA* products can receive downloadable updates of their software. We are constantly working on increasing the compatibility of *STATISTICA* software even with those applications that violate standard

conventions. Therefore, in many circumstances, downloading an update can help when the problem that you are experiencing is caused by nonstandard system configurations or conflicts with other applications.

E-mail technical support. If your question is not answered in the above locations, you may send e-mail to us. Please include your serial number (select ***About STATISTICA*** from the ***Help*** menu to view your serial number) and information about your hardware [the type of processor (CPU) and the amount of memory (RAM) and disk space] and the version of the operating system that you are using.

If you live in North America, send your e-mail to info@statsoft.com; otherwise, e-mail your local StatSoft office (see below).

Phone technical support. You can also call your local StatSoft office to talk to a technician. If you live in the United States or Canada, call (918) 749-1119 (the North American technical support office hours are 9:00 AM to 5:00 PM Central Time, Monday through Friday).

If you live overseas, please contact the office that serves your specific area. You can locate the respective office by selecting ***About STATISTICA*** from the ***Help*** menu and then selecting the ***International Offices*** tab on the ***About STATISTICA*** dialog:

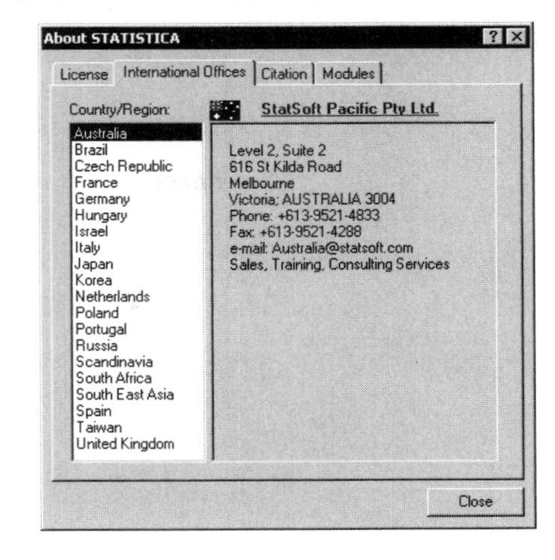

Please have your serial number (select ***About STATISTICA*** from the ***Help*** menu to view your serial number), information about your hardware [the type of processor (CPU) and the amount of memory (RAM) and disk space], and the version of the operating system that you are using ready before you contact StatSoft technical support offices.

APPENDIX

WebSTATISTICA

APPENDIX

WebSTATISTICA

General Overview

Overview (plain language).
WebSTATISTICA Server adds full Internet enablement to *STATISTICA,* including the ability to interactively run *STATISTICA* from a Web browser. With *WebSTATISTICA*, you can easily and quickly access data and powerful analytical tools from virtually any computer in the world as long as it is connected to the Web. The product is provided with a customizable Internet browser-based user interfaces (in the form of simple-to-navigate and easy-to-use dialogs) enabling you to specify analyses and review results. Also, tools are provided to customize these dialogs and easily set up new user interfaces or to add new functions. For example, a simple dialog with only three buttons can be created in the browser, and clicking each button will run a series of analyses and generate a detailed report. *WebSTATISTICA Server* applications add a new dimension and an endless array of possibilities to the entire line of *STATISTICA* Data Analysis, Data Mining, and Quality Control/Six Sigma software.

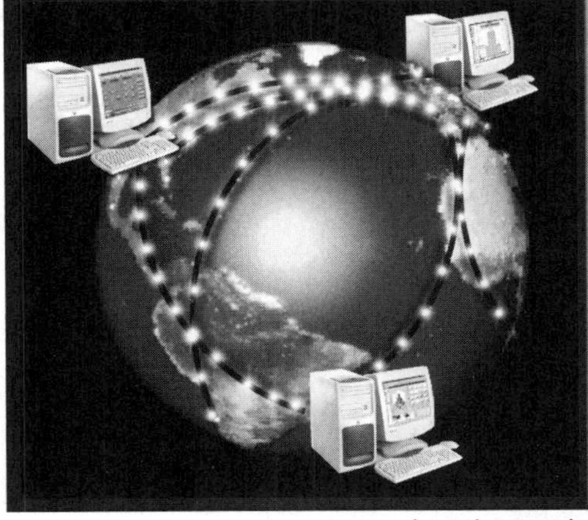

Overview (technical language).
WebSTATISTICA Server is a highly scalable, enterprise-level, Web-based data analysis and database gateway application system that is built on distributed processing technology and fully supports multi-tier Client-Server architecture configurations. *WebSTATISTICA Server* exposes the analytic, query, reporting, and graphics functionality of *STATISTICA* through easy to use, interactive, standard Web interfaces. It is offered as a complete, ready to install application with an interactive, Internet browser-based ("point-and-click") user interface that enables users in remote locations to interactively create data sets, run analyses, and review output. However, *WebSTATISTICA Server* is built using open architecture and includes .NET-compatible development kit tools (based entirely on industry standard syntax conventions such as VB Script, C++, HTML, and XML) that enables IT department personnel to customize all main components of the system or expand it by building on its foundations, for example, by adding new components and/or company-

specific analytic or database facilities. The system is compatible with all major Web server software platforms (e.g., UNIX Apache, and Microsoft IIS), works in both Microsoft .NET and Sun/Java environments, and does not require any changes to the existing firewall and Internet/Intranet security systems.

A Broad Choice of Analytic Facilities and Configurations

The *WebSTATISTICA Server* system is offered as a complete solution that includes the analytic functionality of any *STATISTICA* product or any combination of products, from *STATISTICA Base* to all enterprise systems (*STATISTICA Enterprise* and *Data Miner* applications).

The minimum installation of the *WebSTATISTICA Server* software includes the analytic functionality of *STATISTICA Base* and a license for 5 concurrent users (minimum).

Customers can either order a specific version of *WebSTATISTICA Server* including the analytic functionality that they require (e.g., *STATISTICA Base* for 10 users), or they can add the *Web Server* functionality (as described in this section) to some or all of the seats of the currently licensed *STATISTICA* product (e.g., add the *Web Server* functionality to 20 out of 50 existing licenses of *STATISTICA Enterprise*).

Functionality and Applications: The Advantages of *WebSTATISTICA* Server

A software system that makes *STATISTICA* available "everywhere." Perhaps the clearest advantage offered by the *WebSTATISTICA Server* technology is that it makes the power of any of the *STATISTICA* family of products conveniently available via any computer in the world as long as it is connected to the Internet. Thus, *WebSTATISTICA Server* applications add a new dimension and an endless array of new possibilities and applications to the entire line of *STATISTICA* Data Analysis, Data Mining, and Quality Control/Six Sigma software.

For example, the most recent data and reports (e.g., updated via queries to the specific parts of the corporate data warehouse) – with options to interactively drill down into the results and obtain additional, specific insights about the business – can now be made available to authorized employees wherever they are and regardless of the type of computers to which they have access. Wherever there is the Internet (which means virtually everywhere), there is now also access to the query, reporting, and analytic tools of the most comprehensive data analysis system available.

A powerful, enterprise-wide collaborative-intelligence system. Another, equally important way to take advantage of the *WebSTATISTICA Server* functionality is to use it as the core and natural extension of any of the *STATISTICA* enterprise systems (e.g., *STATISTICA Enterprise* or *Data Miner* applications).

Specifically, *WebSTATISTICA Server* can act as the core of an enterprise-wide network system that enables the participants to work collaboratively and quickly share results (reports), as well as scripts of analyses or queries. User or group permissions (see the Technical Note below) can be used by the administrators to manage access of specific groups of users to specific data or reports. The accessibility of its tools via the Internet makes *WebSTATISTICA Server* a perfect system to facilitate collaborative projects of employees working at different locations or branches of a corporation (even on different continents), or employees who are telecommuting or traveling.

Advantages of distributed processing, and multi-tier Client-Server architecture. Regardless of whether users reside physically close to the location of the *WebSTATISTICA Server* (e.g., in the same building) or far away (e.g., on a different

continent), they will benefit not only from the collaborative work tools but also the options to offload the computationally intensive or time-consuming tasks to the server computers. Specifically, because the most powerful multiprocessor CPUs (and/or multiple computers) are usually used as servers, users can offload computationally intensive tasks, and, for example, run "in the background" queries that will scan terabytes of data on remote servers and perform time consuming long sequences of analyses or reports, while keeping the end users' computers completely free to do other tasks. Because of its distributed processing architecture, *WebSTATISTICA Server* scales in a highly efficient manner to take advantage of multi-processor CPUs and/or multiple computers and, therefore, users can take full advantage of multi-tier Client-Server architecture, where:

- Tier 1 is the user interface on the client computer (a plain browser or *STATISTICA* client, see *STATISTICA Client*, below),
- Tier 2 is the *WebSTATISTICA Server* software and the implementation of the "business intelligence" that it may contain (specific queries, scripts of custom/proprietary analyses, etc.), and
- Tier 3 is *STATISTICA* databases (e.g., *STATISTICA Data Warehouse*) or other corporate repositories of data.

In the desktop version of *STATISTICA,* all computations are performed on the local computer, and resources of other computers are used only in the case when the In-Place Database Processing (IDP) interface to external databases is established. IDP is a technology that reads data asynchronously directly from remote database servers (using distributed processing if supported by the server), and bypasses the need to "import" data and create a local copy of the data set. Records of data are retrieved and sent to the *STATISTICA* computer asynchronously by the CPU of the database server, while *STATISTICA* simultaneously processes them using the CPU of the local computer.

When a Client-Server version of *STATISTICA* is used, the local computer drives only the user interface of *STATISTICA*, and all calculations are performed on the server. The Client-Server architecture offers obvious advantages when your projects are large (e.g., computationally intensive or involving processing of extremely large data sets) and, thus, when they can be offloaded to the servers, freeing your local computer to perform other jobs.

STATISTICA Client. While no components of the *STATISTICA* system are necessary on the client computer (only a browser), having a copy of *STATISTICA* installed on the client side adds new possibilities. One could ask, *"Why would I want to use the WebSTATISTICA Server if I have a copy of STATISTICA installed on my laptop?"* The answer is that having *STATISTICA* installed on the client computer enables you to take additional advantage of the *multi-tier Client-Server* architecture (see above) and work interactively with *STATISTICA* installed locally while offloading certain time-consuming tasks to the server machine(s) and/or exchange data and output between all the three tiers. You can run *WebSTATISTICA* from within desktop *STATISTICA* and flexibly control the interaction between the two. A variety of options are available to share tasks between the desktop and server computer.

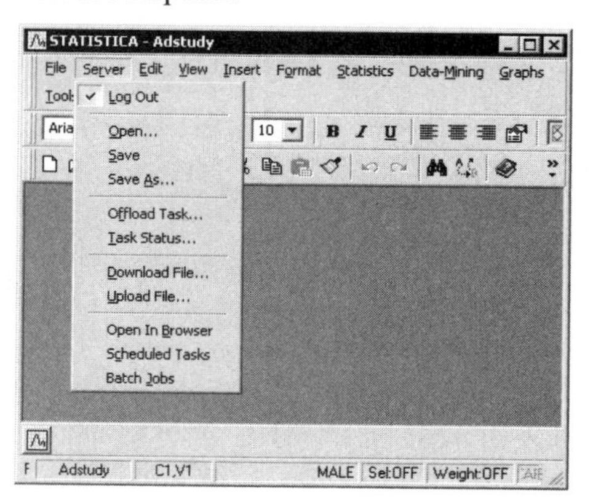

Also, when you review your *WebSTATISTICA* output in the browser, you have options to bring any or all output objects to your desktop computer for further processing. For example, a click on a small button placed optionally (depending on the user configuration) next to every output object (table or graph) sent to your browser by the *WebSTATISTICA Server* system will offer you the option to download that object (a *STATISTICA* table or a graph) to the client computer in its native *STATISTICA* format (in *.sta* or *.stg* file format) so you can work with it offline using the locally installed *STATISTICA* tools.

Advantages of Multithreading Technology

The *WebSTATISTICA* platform is built on advanced distributed processing and multithreading technology to support optimal management of large computational loads. This technology enables rapid processing of even very large and computationally intensive projects, taking full advantage of the multiple CPUs on the server, or even multiple servers working in parallel.

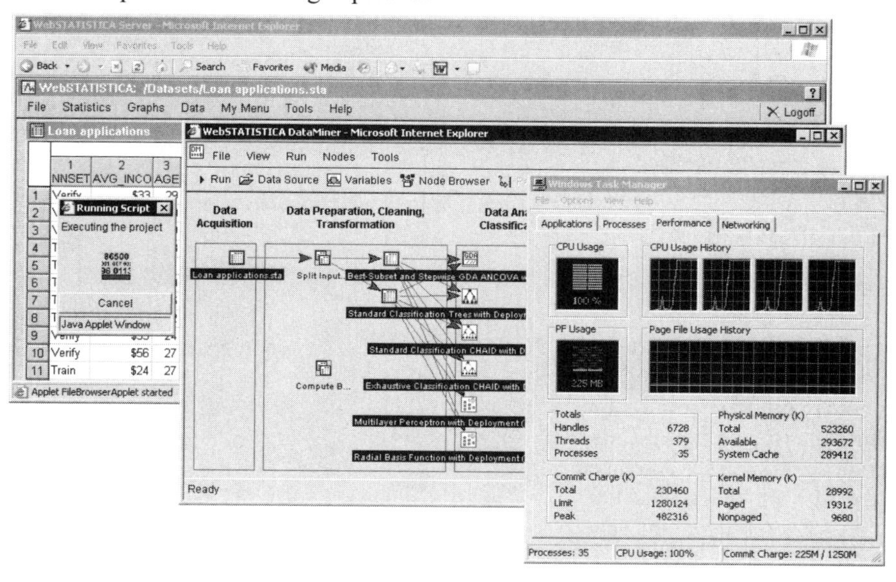

This illustration shows a project running on a quad processor server, along with the server performance monitor demonstrating the full utilization of the resources of all four CPUs executing in the multithreading mode a single, computationally intensive *STATISTICA Data Miner* project.

In addition, the *WebSTATISTICA* architecture delivers a platform-independent, Web browser-based user interface, and provides an ultimate, large enterprise-level ability to manage projects or groups of users "across the hall or across continents."

Ultimate scalability (parallel processing technology). One of the unique features of the *STATISTICA* distributed processing technology is that it flexibly scales not only to take advantage of all CPUs on the current server computer (to support both multiple jobs/users and also individual, computationally intensive projects), but it also scales to multiple server computers. This unique feature is important, since it delivers significant performance gains. *STATISTICA* uses the parallel processing technology across separate hardware units (as some super-computers do) and, therefore, if you have, for example, three servers with four processors each, *STATISTICA* can run an individual project on all 12 processors (if the scale of that project warrants that mode of processing).

WebSTATISTICA User Interface

With the *WebSTATISTICA* implementation of *STATISTICA*, users can interactively run the program from the client machine in a Web browser interface that is similar to that available for the desktop installation. Therefore, the client side of the application (the "front end") can be run on any computer (even a laptop) as long as it is connected to the Internet. However, the actual computations and other operations performed on the data will remain on the (remote) server with its usually more powerful processors and storage resources (and they will be managed using the optimized, multithreading and distributed processing architecture of the system for maximum performance).

In essence, the user interface aspects of *STATISTICA* can be run by one or multiple users from any computer in the world (as long as they are connected to the Internet, even by a slow connection), while the server performs all computations and data operations, enforcing the proper security and access privileges applicable to the respective projects and classes of users, as designed by the network administrator.

WebSTATISTICA offers a straightforward user interface supporting a selection of interactive data analysis, data mining, quality control, database management, database query, and graph customization operations.

After logging on to the *WebSTATISTICA* system,

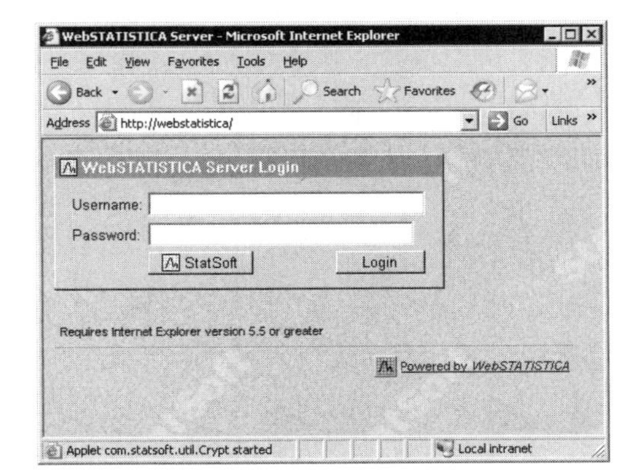

you can select a data source (a data set or a live database connection),

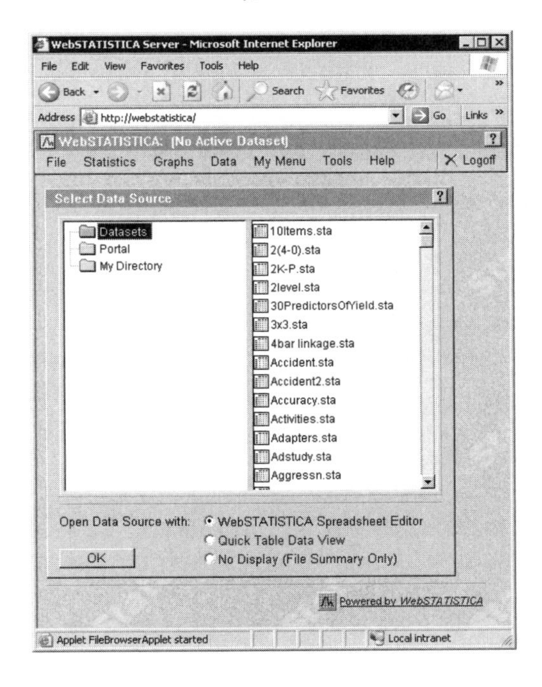

review and edit the data in the interactive *WebSTATISTICA* Spreadsheet Editor,

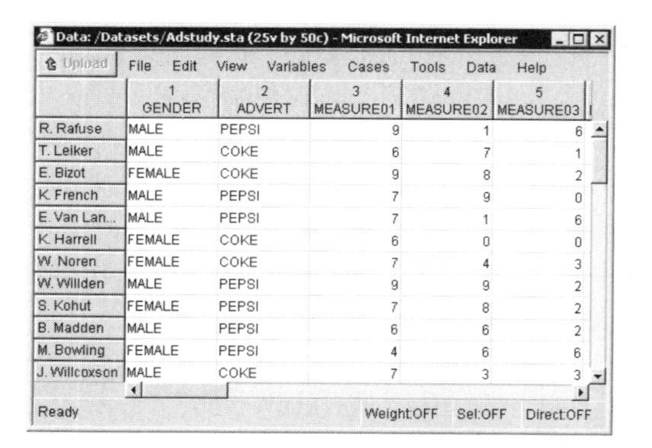

select the analysis to be performed using the standard menu system (or a shortcut in the user-defined ***My Menu***),

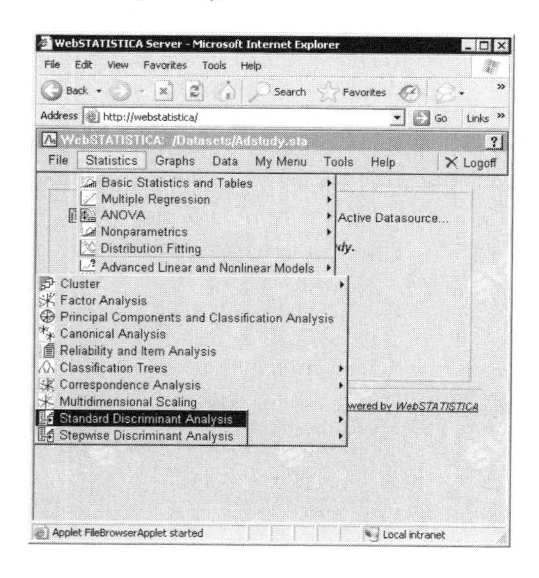

select variables and specify optional analysis parameters,

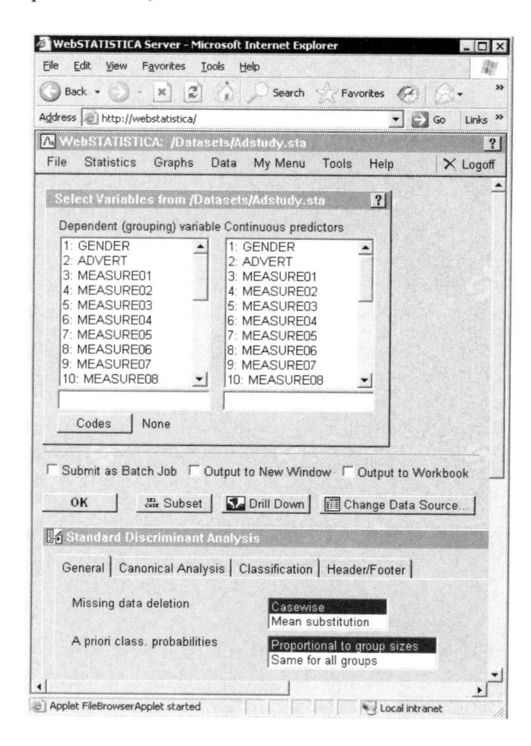

and interactively review the output.

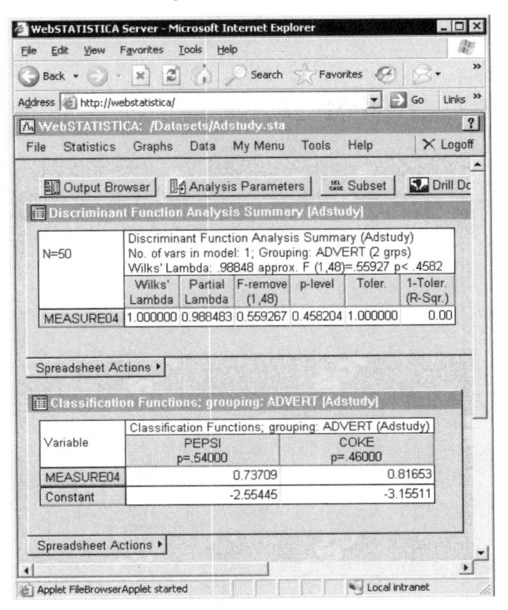

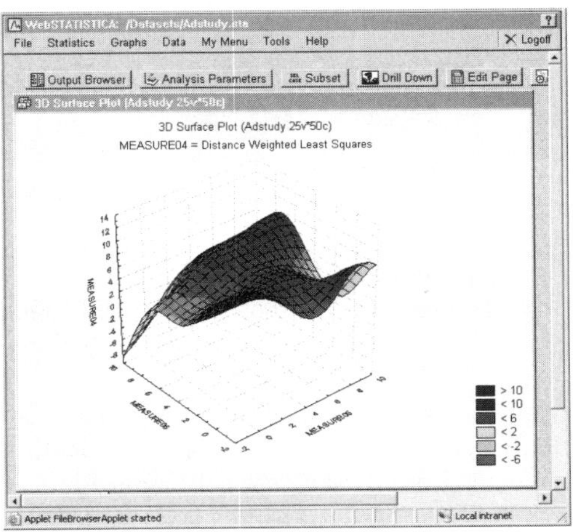

A variety of interactive facilities to perform special database, quality control, or data mining operations (including interactively building data mining models by dragging arrows in the model workspace; see below) are provided, and are accessible interactively from the standard browser.

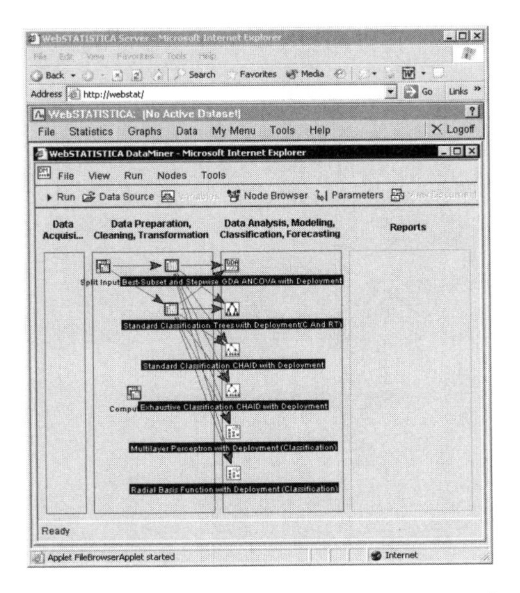

In addition to these built-in, straightforward user-interface facilities, *WebSTATISTICA* also includes a toolkit that enables users to customize the user interface and develop custom applications with specifically predefined functionality, packaged in a way that matches the requirements of their specific applications.

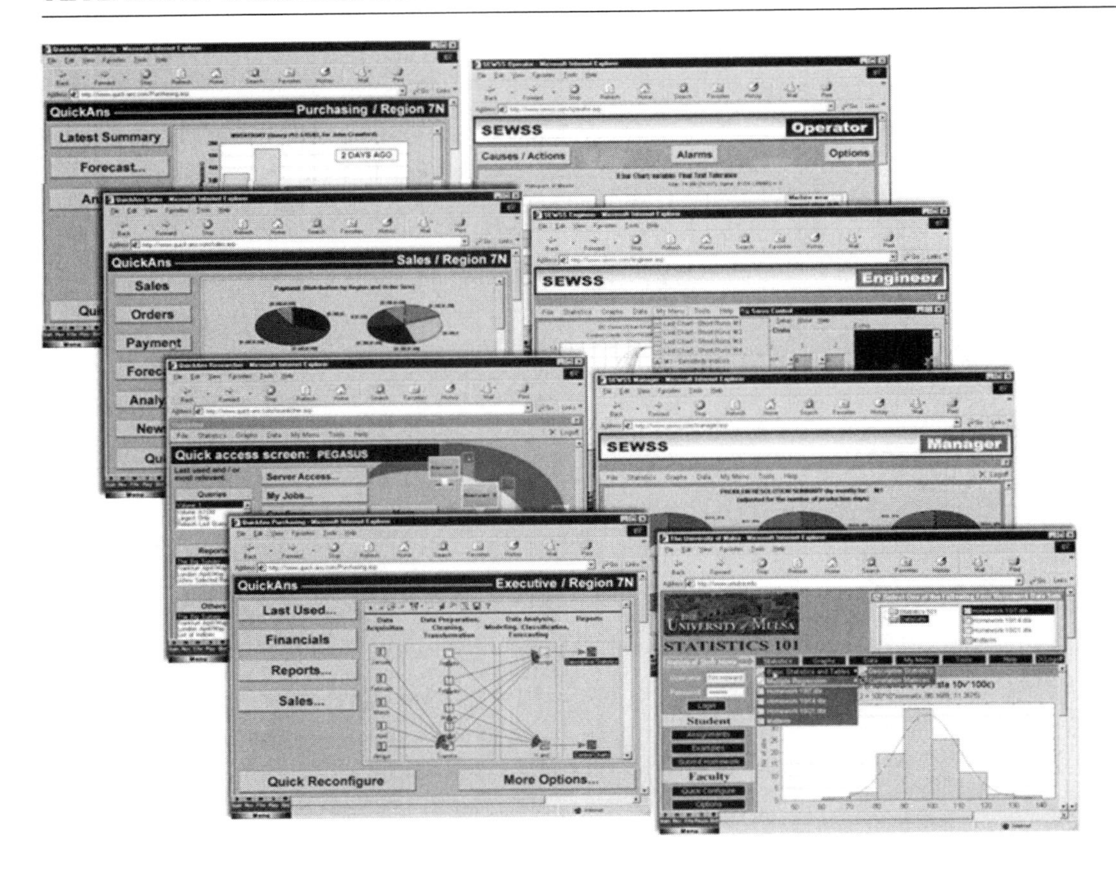

Compatibility with Industry Standards

The unsurpassed compatibility with industry standards is another in the long list of unique advantages of the *WebSTATISTICA Server*.

WebSTATISTICA Server can be deployed on any of the popular Web server platforms (e.g., a UNIX-based Apache or IIS), and therefore, it will conform to the existing local security protocols (fire walls) as required by the corporate client.

WebSTATISTICA Server uses advanced proprietary technology developed at StatSoft to ensure its high performance and scalability (e.g., up to multiple, multiprocessor *STATISTICA Server* computers working in a distributed processing environment). This technology is built on StatSoft's years of experience providing high performance, scalable enterprise systems to major corporations in the United States and around the world. However, *WebSTATISTICA Server* is still based on the industry standard communication protocols (e.g., XML) to ensure (1) its platform independence, (2) smooth transition to future technologies, and (3) ease of customization by the client. Note that the ease of customization is additionally boosted by the fact that only the industry standard syntax conventions (such as VB script, C++, HTML, and XML) are used to customize, configure, and define all the specific analytic operations and all output in *WebSTATISTICA*.

Architecture of the System (A Technical Note)

Although the general design uses two computers in a typical configuration, the Web server (e.g., a UNIX-based Apache system) and at least one *STATISTICA Server* (optionally scalable to multiple *STATISTICA Servers*),

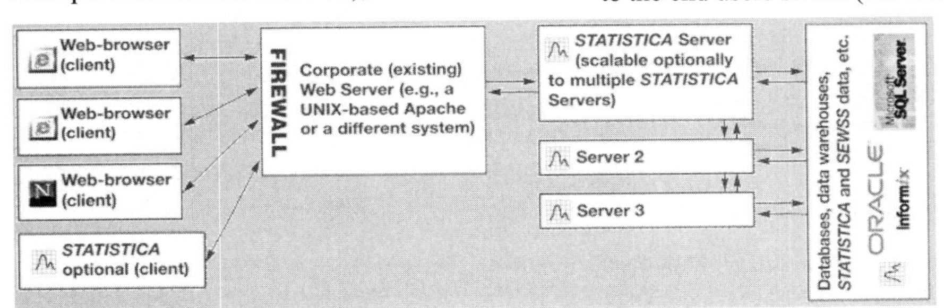

in many cases, the *STATISTICA Server* could be installed on the same machine if desired (when IIS is used as the Web host):

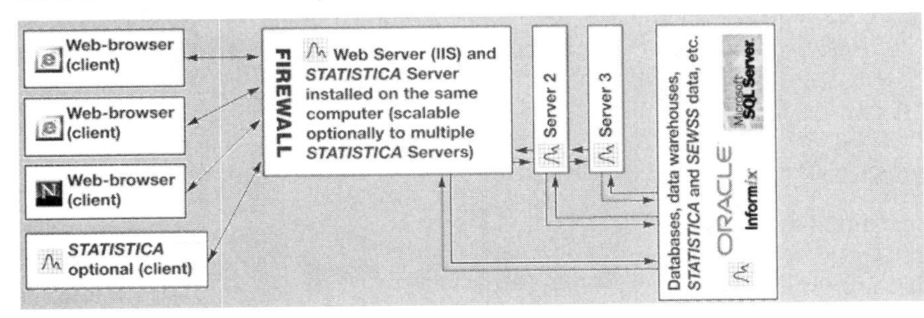

The design allows for a flexible, generic Web server implementation by using a standard scripting language on the Web server. The purpose of the Web server is to package requests from the user (received from a browser), send these to the *STATISTICA Server*, and then process responses from the *STATISTICA Server* for display to the users (on their browsers).

Communication between the Web server and the *STATISTICA Server* is accomplished through technology based on the industry standard XML conventions. The system is fully customizable, and for customers who want to develop their own modifications or extensions of this (ready to deploy) system, it provides development tool kit facilities allowing modification of all aspects of both the scripts that are being executed by *STATISTICA* (on the *STATISTICA Server* side) and the appearance of the user interface exposed to the end users on the (browser-based) thin client side. Only the most standard, commonly known tools (such as VB or XML/HTML) are used to customize or expand the system.

The actual Web page definitions and *STATISTICA* scripts to be executed are stored in a designated Repository Facility on the *STATISTICA Server*, and they are managed in a queue-like fashion. The system also includes a highly optimized Distributed Processing Manager that handles the incoming processing load and distributes it optimally over multiple threads of *STATISTICA* and multiple *STATISTICA Server* computers.

The *WebSTATISTICA Server* software system also includes the *STATISTICA* Visual Basic Web Extensions. These extensions to the SVB language enable the script writer to either let the system display the resulting graphs and spreadsheets on the automatically generated

(output) Web pages, or customize the appearance of the generated output pages by adding HTML directives as appropriate.

Security and authentication is a key design feature in the *WebSTATISTICA Server* application system. At the beginning of the session, users "sign on" to the system with their user name and password. System administrators are able to control access to data sources and scripts based either on user or group permissions. The highest level of the access privilege allows advanced users (or administrators) to execute virtually arbitrary scripts (e.g., in order to perform system administration or maintenance operations). This level requires a designated (highest) access privilege because, due to the general nature and power of the *STATISTICA* Visual Basic language, it gives access (to the authorized users) to all resources on the network.

Note that this system can be integrated with the "traditional" (i.e., non-Web-based) *STATISTICA* concurrent network or a *STATISTICA* enterprise system authentication scheme so that a corporate customer can install, for example, a 50-user (total) *STATISTICA* enterprise system or a concurrent network with 20 of them accessible via the *WebSTATISTICA Server*.

Competitive Advantages

The competitive advantages of the *WebSTATISTICA Server* applications start obviously with the complete list of unique features of *STATISTICA* itself. Further, unlike the competing products, we offer a complete application (a "solution") with a Web-based user interface and not merely a "development kit" (although the development kit facilities are also available to extend or customize the system). Also, we do not require that a specific Web server software be installed first (which may or may not comply with the client's security standards and other policies). Finally, our system is controlled by industry standard VB scripts, C++, HTML, and XML that can be easily modified by users or system administrators. Also, our distributed processing and multithreading technology delivers performance and system responsiveness that is not matched by any competing products.

WebSTATISTICA Knowledge Portal

A designated *WebSTATISTICA Knowledge Portal* application is optionally available that enables users to effectively and securely distribute organized sets of output documents over the Web. It offers support for workgroups of users (each with different access privileges, and thus access to different parts of the database of output documents), intuitive tree-view organization of available materials, and options to broadcast documents updated on the Web server in real time.

WebSTATISTICA Demo Movie

How does *WebSTATISTICA Server* work? Visit StatSoft's Web site, www.statsoft.com, to view an informative presentation of the unique features of *WebSTATISTICA* described here. The movie also includes a step-by-step example application.

APPENDIX

STATISTICA FAMILY OF PRODUCTS

continued ➲

APPENDIX

STATISTICA FAMILY OF PRODUCTS

Common system features. In addition to comprehensive, leading edge analytics, *STATISTICA* products offer a selection of fully customizable user interfaces (with simplified shortcut templates for novices), flexible, presentation-quality output management (including a variety of report formats, such as *.pdf*, Word, *.rtf*, *.html*, and output to Web portals), full OLE/ActiveX support, and Web enablement.

Also, all products include data management optimized to handle large data sets, interactive database query tools, and a wide selection of data import/export facilities. *STATISTICA* products can handle data sets of practically unlimited size and offer "quadruple" precision calculations; they support multiple input files, multiple instances, and multitasking. A broad selection of interactive visualization and graphics/drawing tools of the highest quality is fully integrated into each product, and each includes a complete set of automation options and a professional Visual Basic and .NET-compatible development environment with more than 13,000 externally accessible functions.

GENERAL-PURPOSE DESKTOP PRODUCTS

***STATISTICA* Base.** Offers a comprehensive set of essential statistics in a user-friendly package and all the performance, power, and ease of use of the *STATISTICA* technology.

- All *STATISTICA* graphics tools
- Basic Statistics, Breakdowns, and Tables
- Distribution Fitting
- Multiple Linear Regression
- Analysis of Variance
- Nonparametrics, and more.

***STATISTICA* Advanced Linear/Nonlinear Models.** Offers a wide array of the most advanced modeling and forecasting tools on the market, including automatic model selection facilities and extensive interactive visualization tools.

- General Linear Models
- Generalized Linear/Nonlinear Models
- General Regression Models
- General Partial Least Squares Models
- Variance Components
- Survival Analysis
- Nonlinear Estimation

- Fixed Nonlinear Regression
- Log-Linear Analysis of Frequency Tables
- Time Series/Forecasting
- Structural Equation Modeling, and more.

 ***STATISTICA* Multivariate Exploratory Techniques.** Offers a broad selection of exploratory techniques for various types of data, with extensive, interactive visualization tools.

- Cluster Analysis
- Factor Analysis
- Principal Components/Classification Analysis
- Canonical Analysis
- Discriminant Analysis
- General Discriminant Analysis Models
- Reliability/Item Analysis
- Classification Trees
- Correspondence Analysis
- Multidimensional Scaling, and more.

 ***STATISTICA* Variance Estimation and Precision (VEPAC).** A comprehensive set of techniques for analyzing data from experiments that include both fixed and random effects using REML (Restricted Maximum Likelihood Estimation). With *STATISTICA VEPAC*, you can obtain estimates of variance components and use them to make precision statements while at the same time comparing fixed effects in the presence of multiple sources of variation.

- Variability plots
- Multiple plot layouts to allow direct comparison of multiple dependent variables
- Expected mean squares and variance components with confidence intervals
- Flexible handling of multiple dependent variables: analyze several variables with the same or different designs at once
- Graph displays of variance components

 ***STATISTICA* Automated Neural Networks (SANN).** Contains the most comprehensive neural network algorithms and training methods.

- Automatic Search for Best Architecture and Network Solutions
- Multilayer Perceptrons
- Radial Basis Function Networks
- Self-Organizing Feature Maps
- Time Series Neural Networks for both Regression and Classification problems
- A variety of algorithms for fast and efficient training of Neural Network Models including Gradient Descent, Conjugate Gradient, and BFGS
- Numerous analytical graphs to aid in generating results and drawing conclusions
- Sampling of data into subsets for optimizing network performance and enhancing the generalization ability
- Sensitivity Analysis, Lift Charts, and ROC Curves
- Creation of Ensembles out of already existing standalone networks
- C-code and PMML (Predictive Model Markup Language) Neural Network Code Generators that are easy to deploy.

 ***STATISTICA* Power Analysis.** An extremely precise and user-friendly, specialized tool for analyzing all aspects of statistical power and sample size calculation.

- Sample Size Calculation
- Confidence Interval Estimation
- Statistical Distribution Calculators, and more.

INDUSTRIAL SOLUTIONS, SIX SIGMA TOOLS

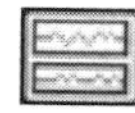 ***STATISTICA* Quality Control Charts.** Offers fully customizable (e.g., callable from other environments), easy and quick to use, versatile charts with a

selection of automation options and user-interface shortcuts to simplify routine work (a comprehensive tool for Six Sigma methods).

- Multiple Chart (Six Sigma Style) Reports and displays
- X-bar and R Charts; X-bar and S Charts; N_p, P, U, C Charts
- Pareto Charts
- Process Capability and Performance Indices
- Moving Average/Range Charts, EWMA Charts
- Short Run Charts (including Nominal and Target)
- CuSum (Cumulative Sum) Charts
- Runs Tests
- Interactive
- Causes and actions, customizable alarms, analytic brushing, and more.

***STATISTICA* Process Analysis.** A comprehensive package for Process Capability, Gage R&R, and other quality control/improvement applications (a comprehensive tool for Six Sigma methods).

- Process/Capability Analysis Charts
- Ishikawa (Cause and Effect) Diagrams
- Gage Repeatability & Reproducibility
- Variance Components for Random Effects
- Weibull Analysis
- Sampling plans, and more.

***STATISTICA* Design of Experiments.** Features the largest selection of DOE and related visualization techniques including interactive desirability profilers (a comprehensive tool for Six Sigma methods).

- Fractional Factorial Designs
- Mixture Designs
- Latin Squares
- Search for Optimal $2^{(k-p)}$ Designs
- Residual Analysis and Transformations
- Optimization of single/multiple response variables
- Central Composite Designs
- Taguchi Designs
- Minimum Aberration & Maximum Unconfounding
- $2^{(k-p)}$ Fractional Factorial Designs with Blocks
- Constrained Surfaces
- D- and A-Optimal Designs
- Desirability profilers, and more

***STATISTICA* Multivariate Statistical Process Control (MSPC).** A complete solution for multivariate statistical process control, deployed within a scalable, secure analytics software platform.

- Univariate and multivariate statistical methods for quality control, predictive modeling, and data reduction
- Functions to determine the most critical process, raw materials, and environment factors and their optimal settings for delivering products of the highest quality
- Monitoring of process characteristics interactively or automatically during production stages
- Building, evaluating, and deploying predictive models based on the known outcomes from historical data
- Historical analysis, data exploration, data visualization, predictive model building and evaluation, model deployment to monitoring server
- Interactive monitoring with dashboard summary displays and automatic-updating results
- Automated monitoring with rules, alarm events, and configurable actions
- Multivariate techniques including Partial Least Squares, Principal Components, Neural Networks, Recursive Partitioning (Tree) Methods, Support Vector Machines, Independent Components Analysis, Cluster Analysis, and more

STATISTICA ENTERPRISE SYSTEMS

In addition to the common features listed above, *STATISTICA* Enterprise Systems optionally offer a wide selection of tools for collaborative work, Web browser-based user interfaces (using *WebSTATISTICA Server*), specialized databases, and a highly optimized interface to enterprise-wide data repositories, including options to rapidly process large data sets from remote servers in-place, without creating local copies. Deployment and on-site training services are available.

***STATISTICA* Data Miner.** The most comprehensive selection of data mining solutions on the market, with an icon-based, extremely easy to use user interface (optionally Web browser based via *WebSTATISTICA*, see page 286) and a deployment engine. It features a selection of completely integrated and automated, ready to deploy "as is" (but also easily customizable) systems of specific data mining solutions for a wide variety of business applications. A designated SPC version (*QC Data Miner*) to mine/analyze large streams of QC data is also available. The data mining solutions are driven by powerful procedures from five modules:

- General Slicer/Dicer Explorer (with optional OLAP)
- General Classifier
- General Modeler/Multivariate Explorer
- General Forecaster
- General Neural Networks Explorer, and more

***STATISTICA* QC Miner.** A powerful software solution designed to monitor processes and identify and anticipate problems related to quality control and improvement with unmatched sensitivity and effectiveness. *STATISTICA QC Miner* integrates all Quality Control Charts, Process Capability Analyses, Experimental Design procedures, and Six Sigma methods with a comprehensive library of cutting-edge techniques for exploratory and predictive data mining.

- Predict QC problems with cutting edge data mining methods
- Discover root causes of problem areas
- Monitor and improve ROI (Return On Investment)
- Generate suggestions for improvement
- Monitor processes in real time over the Web
- Create and deploy QC/SPC solutions over the Web
- Use multithreading and distributed processing to rapidly process extremely large streams of data

***STATISTICA* Text Miner.** A powerful software solution for text mining, document retrieval, and mining of unstructured data. An optional add-on product for *STATISTICA Data Miner,* designed and optimized for accessing and analyzing documents (unstructured information) in a variety of formats: *.txt* (text), *.pdf* (Adobe), *.ps* (PostScript™), *.html*, *. xml* (Web-formats), and most Microsoft Office formats (e.g., *.doc, .rtf*); optimized access to Web pages (URL addresses) is also provided.

- Efficiently index very large collections of text documents; identify key terms and similarities between documents and terms, and extract the information relevant to your mission and goals
- Apply stub-lists (words to ignore) and language-specific stemming algorithms (various languages are supported)
- Includes numerous options for converting documents into numeric information for further processing (e.g., mapping, clustering, predictive data mining, classification of documents, etc.)
- Full support for multithreaded operation on multi-processor server installations for

extremely fast indexing and searching of huge document repositories

- Can also be used to index, analyze, and mine other unstructured input, such as sound or image files (after domain-specific pre-processing is applied)
- Fully integrated into the *STATISTICA* and *WebSTATISTICA* systems; hence, the large number of available methods for supervised and unsupervised learning (clustering), mapping, data visualization, etc., are directly and immediately available; many of the algorithms available in *STATISTICA Data Miner*, such as the machine learning algorithms (*k*-Nearest Neighbor, Naive Bayes classifiers, advanced Support Vector Machines and Kernel classifiers), are particularly well suited for text mining or the analysis of other unstructured information

STATISTICA Sequence, Association and Link Analysis (SAL). Designed to address the needs of clients in retailing, banking, insurance, etc., industries by implementing the fastest known highly scalable algorithm with the ability to drive Association and Sequence rules in one single analysis. The program represents a stand-alone module that can be used for both model building and deployment. All tools in *STATISTICA Data Miner* can be quickly and effortlessly leveraged to analyze and "drill into" results generated via *STATISTICA SAL*.

- Uses a tree-building technique to extract Association and Sequence rules from data
- Uses efficient and thread-safe local relational database technology to store Association and Sequence models
- Handles multiple response, multiple dichotomy, and continuous variables in one analysis
- Performs Sequence Analysis while mining for Association rules in a single analysis
- Simultaneously extracts Association and Sequence rules for more than one dimension
- Given the ability to perform multidimensional Association and Sequence mining and the capacity to extract only rules for specific items, the program can be used for Predictive Data Mining
- Performs Hierarchical Single-Linkage Cluster Analysis, which can detect the more likely cluster of items that can occur. This has extremely useful, practical real-world applications, e.g., in retailing.

STATISTICA Enterprise. An integrated multi-user system designed for general-purpose data analysis and business intelligence applications in research. *STATISTICA Enterprise* can optionally offer the statistical functionality available in any or all *STATISTICA* products.

- Integration with data warehouses
- Intuitive query and filtering tools
- Easy-to use administration tools
- Automatic report distribution
- Alarm notification, and more

STATISTICA Enterprise/QC. Designed for local and global enterprise quality control and improvement applications including Six Sigma. *STATISTICA Enterprise/QC* offers a high-performance database (or an optimized interface to existing databases), real-time monitoring and alarm notification for the production floor, a comprehensive set of analytical tools for engineers, sophisticated reporting features for management, Six Sigma reporting options, and much more.

- Web-enabled user interface and reporting tools; interactive querying tools
- User-specific interfaces for operators, engineers, etc.
- Groupware functionality for sharing queries, special applications, etc.

- Open-ended alarm notification including cause/action prompts
- Scalable, customizable, and can be integrated into existing database/ERP systems, and more

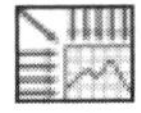 ***STATISTICA* Monitoring and Alerting Server (MAS).** A system that enables users to automate the continual monitoring of hundreds or thousands of critical process and product parameters. The ongoing monitoring is an automated and efficient method for:

- Monitoring many critical parameters simultaneously
- Providing status "snapshots" from the results of these monitoring activities to personnel based on their responsibilities.
- Dashboards associated with User/Group.

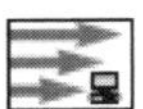 ***STATISTICA* MultiStream.** A solution package for identifying and implementing effective strategies for advanced multivariate process monitoring and control. *STATISTICA MultiStream* was designed for process industries in general, but is particularly well suited to help power generation facilities leverage their data (collected into existing specialized process data bases for multivariate and predictive process control) for actionable advisory systems.

STATISTICA MultiStream is a complete enterprise system built on a robust, advanced client-server (and fully Web-enabled) architecture, offers central administration and management of deployment of models, as well as cutting edge root-cause analysis and predictive data mining technology, and its analytics are seamlessly integrated with a built-in document management system.

- Automated (nonlinear) root cause analysis and feature selection for thousands of parameters, to clearly identify which ones are the most likely responsible for process problems
- Automated and interactive commonality analysis to identify parameters and processes that shifted or moved from normal operations during particular time intervals
- Advanced linear and nonlinear (e.g., SVM, Recursive Partitioning, Neural Nets) models for creating sensitive multivariate control schemes and work flows to identify multivariate shifts and drifts early, before they cause problems
- Advanced data mining algorithms for predicting and optimizing key performance and quality indicators
- Tracks hundreds of data streams simultaneously
- Delivers simple summaries relevant to critical process parameters and outcomes via efficient and simple dashboards and drill-down workflows
- Delivers standard and customized analytic workflows for root cause analysis, leveraging cutting-edge data analysis and data mining technologies
- Warns of (predicted) problems and equipment failures before they occur (predictive alarming), thus avoiding costly shut-downs and unscheduled maintenance
- Watches "everything" that impacts your process performance in real time

 ***WebSTATISTICA* Server.** The ultimate enterprise system that offers full Web enablement, including the ability to run *STATISTICA* interactively or in batch from a Web browser on any computer (including Linux, UNIX), offload time consuming tasks to the servers (using distributed processing), use multi-tier Client-Server architecture, manage projects over the Web, and collaborate "across the hall or across continents" (supporting multithreading and distributed/parallel processing that scales to multiple server computers).

DATA AND DOCUMENT MANAGEMENT

***STATISTICA* Document Management System (SDMS).** A complete, highly scalable, database solution package for managing electronic documents. With the *STATISTICA Document Management System*, you can quickly, efficiently, and securely manage documents of any type [e.g., find them, access them, search for content, review, organize, edit (with trail logging and versioning), approve, etc.].

- Extremely transparent and easy to use
- Flexible, customizable (optionally browser/Web-enabled) user interface
- Electronic signatures
- Comprehensive auditing trails, approvals
- Optimized searches
- Document comparison tools
- Security
- Satisfies the FDA 21 CFR Part 11 requirements
- Satisfies ISO 9000 (9001, 14001) documentation requirements
- Unlimited scalability (from desktop or network Client-Server versions, to the ultimate size, Web-based worldwide systems)
- Open architecture and compatibility with industry standards

***STATISTICA* PI Connector.** An optional *STATISTICA* add-on component that allows for direct integration to data stored in the PI data historian. The *STATISTICA PI Connector* utilizes the PI user access control and security model, allows for interactive browsing of tags, and takes advantages of dedicated PI functionality for interpolation and snapshot data. *STATISTICA* integrated with the PI system is being used for streamlined and automated analyses for applications such as Process Analytical Technology (PAT) in FDA-regulated industries, Advanced Process Control (APC) systems in Chemical and Petrochemical industries, and advisory systems for process optimization and compliance in the Energy Utility industry.

***STATISTICA* Data Warehouse.** A complete, powerful, scalable, and customizable intelligent data warehouse solution, which also optionally offers the most complete analytic functionality available on the market, fully integrated into the system. *STATISTICA Data Warehouse* consists of a suite of powerful, flexible component applications, including:

- *STATISTICA* Data Warehouse Server Database
- *STATISTICA* Data Warehouse Query (featuring *WebSTATISTICA Query*)
- *STATISTICA* Data Warehouse Analyzer (featuring *WebSTATISTICA Data Miner*, *WebSTATISTICA Text Miner*, *WebSTATISTICA QC Miner*, or the complete set of *WebSTATISTICA* analytics)
- *STATISTICA* Data Warehouse Reporter (featuring *WebSTATISTICA Knowledge Portal* and/or *WebSTATISTICA Interactive Knowledge Portal*)
- *STATISTICA* Data Warehouse Document Repository (featuring *WebSTATISTICA Document Management System*)
- *STATISTICA* Data Warehouse Scheduler
- *STATISTICA* Data Warehouse Real Time Monitor and Reporter (featuring *WebSTATISTICA Enterprise* or *WebSTATISTICA Enterprise/QC*)

If you are new to data warehousing, StatSoft consultants will guide you step by step through the entire process of designing the optimal data warehouse architecture – from a comprehensive review of your information storage and extraction/analysis needs, to the

final training of your employees and support of your daily operations.

Crucial features and benefits. The crucial features and benefits of *STATISTICA Data Warehouse* solutions include, among many others:

- Complete data warehousing application tailored to your business
- Platform independent architecture for seamless integration with your existing infrastructure
- Facilities to integrate data from a wide variety of sources
- Virtually unlimited scalability
- Options to update/synchronize data from multiple sources via automatic schedulers or on demand
- Completely Web-enabled system architecture to provide ultimate enterprise functionality for all company locations around the world (e.g., access via Web browsers from any location)
- Advanced security model and authentication of users
- Complete document management options to optimize management of documents of any types and satisfy regulatory requirements (e.g., FDA 21 CFR Part 11, ISO 9000)
- Advanced analytic components to clean/verify data and to integrate automated data mining, artificial intelligence, and real-time process monitoring
- Options to automatically run and post on Knowledge Portals (or broadcast) highly customized reports, including interactive (i.e., drillable, sliceable, and user-customizable) reports and results of advanced analytics
- Backup and archiving options
- Programmable, customizable, and expandable to adapt to specific mission profiles (open architecture, exposed to extensions using the most industry standard languages, such as VB, C++, Java, HTML)
- Built on robust, well tested, highly scalable, cutting-edge technology to leverage your investment [including highly optimized in-place database processing (IDP) technology, true multithreading, distributed/parallel processing, and support for pooling CPU resources of multiple servers to deliver supercomputer-like performance]

STATISTICA Data Warehouse is a complete intelligent data storage and information delivery/distribution solution that enables you to customize the flow of information through your organization, provide all authorized members of your organization with flexible, secure, and rapid access to critical information and intelligent reporting.

The system is virtually platform independent and will fit into any existing database architecture and hardware environment. It will efficiently combine information from multiple database formats and sources (from manual data entry forms to large batteries of automatic data collection devices). The system can be further enhanced through integration with other fully compatible components of the *STATISTICA* line of applications and solutions; to name just a few:

- *STATISTICA Data Miner* for advanced data mining and artificial intelligence (e.g., neural networks) based solutions to provide decision support through cutting-edge methods for knowledge extraction and prediction
- *Quality Control Miner* and *Enterprise/QC* for tight integration with quality control, process control, and yield management activities
- *STATISTICA Text Miner* for automatic processing of unstructured information in documents, databases, or Web directories (Web-crawling of URLs)
- *STATISTICA Knowledge Portal* for presenting summary reports, charts, and action items to end users (management, sales force, engineers, etc.) through secure access portals via the Web; to deliver key

intelligence and decision support to stakeholders worldwide (e.g., you access the *STATISTICA Knowledge Portal* via standard Web browsers from anywhere in the world)

Architecture and connectivity. *STATISTICA Data Warehouse* connects to any platform, database, or data source, and will scale to businesses and applications of any size. The program is built on a database and database schema customized for your particular business. The solution can be installed either inclusive of a high performance database engine (SQL Server) or as a (virtual) database schema compatible with most industry standard databases; therefore, it will seamlessly integrate into existing database systems.

Because *STATISTICA Data Warehouse* does not depend on one particular database vendor or hardware platform, it is itself entirely platform-independent. The main *Data Warehouse* software will connect to any database format and, hence, can efficiently combine and pool information from multiple sources.

STATISTICA Data Warehouse application software will run on servers with multiple processors or banks of multiple-processor servers for super-computer like performance. The system will scale effortlessly and economically to even huge data sizes and analysis (intelligence) problems.

Web enablement. *STATISTICA Data Warehouse* extracts information from sources anywhere in the world and delivers intelligence anywhere in the world.

The Web component of the system is built on the proven *WebSTATISTICA* technology that is used by organizations worldwide to provide secure access via standard Web browsers. Unlike other Web-based solutions, *STATISTICA Data Warehouse* does not require any additional components to be installed on the (thin) client machines. Hence, the system can be utilized by (authorized and authenticated) users worldwide from hotel rooms via dial-up modems, from home, or from office and production facilities located at the most remote places on earth (e.g., via satellite Web links).

Advanced security and authentication. The *STATISTICA Data Warehouse* implements a detailed and sophisticated security system to ensure that your proprietary knowledge and intelligence is safe from unauthorized access. The system will likely become the most important repository of business intelligence and decision support resources in your organization. Therefore, the security of the system is a crucial priority so that those valuable resources are shielded from unauthorized access.

STATISTICA Data Warehouse implements the highest level of security by establishing groups of users with different levels of authority (regarding the information that is accessible and the operations that can be performed), requiring regularly updated passwords, etc. Special methods are also in place to detect and guard against systematic electronic intrusions ("hacking").

Document control. *STATISTICA Data Warehouse* enables full document management, compliant with government and industry standards.

STATISTICA Document Management System can be seamlessly integrated into your *STATISTICA Data Warehouse* application to optimize the flow of information within your organization and thus increase your productivity. This system can also be configured to comply with all (corporate) documentation management policies or regulatory requirements for document security, audit trails, and electronic signatures/authentication (as, for example, stipulated by FDA 21 CFR Part 11: *Electronic Records; Electronic Signatures*; or ISO 9001 4.5: *Document and data control*).

Advanced analytics. *STATISTICA Data Warehouse* can incorporate the most advanced data analysis and knowledge extraction methods available; you can go far beyond OLAP to simplify and extract knowledge about even the most complex – and inaccessible to other applications – patterns in the data.

Because *STATISTICA Data Warehouse* is built from the same high performance components as the entire *STATISTICA* line of analytic solutions software, those analytic solutions can easily and seamlessly be integrated into your Data Warehouse. *STATISTICA* offers the most comprehensive set of tools for data mining, text mining, data analysis, graphics and visualization, quality and process control (including Six Sigma), etc. on the market. These resources and technologies can be connected to the data sources in the *STATISTICA Data Warehouse* to leverage the most advanced technologies and algorithms available for analyzing and extracting key intelligence from all sources. For example, you can apply hundreds of neural networks architectures, highest performance tree classifiers (e.g., stochastic gradient boosting trees), flexible root cause analyses, control charting methods, powerful business forecasting methods, or sophisticated analytic graphics methods to convert raw data in the Data Warehouse into useful and actionable intelligence with clear implications for decisions affecting your business.

Programmability and customizability. *STATISTICA Data Warehouse* is an open-architecture system that will not lock you into a relation with a single vendor or solution; you can respond quickly to new business demands and requirements that need to be incorporated into the Data Warehouse.

As all applications and solutions in the *STATISTICA* family of products, *STATISTICA Data Warehouse* is fully programmable and customizable, using industry standard programming tools such as Visual Basic, C++, Java, or HTML. This feature is of key importance when your business depends on your ability to quickly adapt to new information and business realities. Because you can customize the system without being forced to rely on the programmers of a single vendor or knowledge of idiosyncratic scripting conventions (required by many competing solutions), you have the freedom to develop your proprietary extensions to the data warehouse and to add not only your own reports but also custom analytic and data transformation/cleaning procedures, using widely available resources and industry standard tools (e.g., VB, C++, Java, or HTML tools and programmers). Of course, StatSoft can always offer to you a full complement of consulting, system integration, and programming services delivered by an experienced staff, if you choose to work with us.

VERTICAL MARKET APPLICATIONS

PROCEED. A turnkey manufacturing software solution that distills fundamental causal relationships between products and the processes that produce them, using data that is already collected and managed. *PROCEED* implements the patent-pending approach developed and proven at Caterpillar Inc. and powered by the *STATISTICA* Enterprise Analytics Software Platform.

High tech manufacturing enterprises today collect vast amounts of data.

- Data about the production processes.
- Data about tests of raw materials, subassemblies, and materials in process.
- Data about the critical to quality attributes of finished products.

All of these data collection and storage efforts continue to be fueled by increases in automation, technology advances in the storage capabilities of data repositories, and the advances in sensors and other techniques for measurement. Today's manufacturers are sitting on a gold mine of information . . . only if they are able to translate it into actionable information.

Collecting data is not sufficient to drive enterprise change. To create change, we need to translate these data into knowledge and then communicate that knowledge in a format that enables the people who are empowered to act on it. Now is the time for this Return on Investment from data using *PROCEED*.

PROCEED combines novel and traditional knowledge extraction methods to:

- Derive and validate simple to complex causal relationships between manufacturing processes and product quality outcomes
- Deploy actionable information to enable process owners and knowledge workers to compare what-if scenarios and simultaneously optimize multiple competing outcomes

***STATISTICA* *PowerSolutions*.** A solution package aimed for use at power generation companies to optimize power plant performance, increase efficiency, and reduce emissions. This product offers a highly economical alternative to multimillion dollar investments in new or upgraded equipment (hardware). Based on more than 20 years of experience in applying advanced data driven, predictive data mining/optimization technologies for process optimization in various industries, *STATISTICA PowerSolutions* enables power plants to get the most out of their existing equipment and control systems by leveraging all data collected at their sites to identify opportunities for improvement, even for older designs such as coal-fired Cyclone furnaces (as well as wall-fired or T-fired designs).

INDEX

A

B

C

D

E

StatSoft

STATISTICA

StatSoft